EUROPEAN AIR TRAFFIC MA

European Air Traffic Management
Management
Principles, Practice and Research

Edited by

ANDREW COOK
University of Westminster, UK

Routledge
Taylor & Francis Group

LONDON AND NEW YORK

First published 2007 by Ashgate Publishing

2 Park Square, Milton Park, Abingdon, Oxon OX14 4RN
711 Third Avenue, New York, NY 10017, USA

Routledge is an imprint of the Taylor & Francis Group, an informa business

First issued in paperback 2016

British Library Cataloguing in Publication Data
European air traffic management : principles, practice and
    research
    1. Aeronautics, Commercial - Europe - Management 2. Air
    traffic control - Europe 3. Air traffic capacity - Europe
    I. Cook, Andrew
    387.7'4042'094

The Library of Congress has cataloged the printed edition as follows:
European air traffic management : principles, practice, and research / [compiled] by
Andrew Cook.
        p. cm.
    Includes bibliographical references and index.
    ISBN: 978-0-7546-7295-1
    1. Air traffic control--Europe. I. Cook, Andrew, Dr.

    TL725.3.T7E94 2008
    387.7'40426094--dc22

                                                                    2007030988

ISBN: 978-0-7546-7295-1 (hbk)
ISBN: 978-1-138-25576-0 (pbk)

Contents

List of Figures

List of Tables

List of Contributors

Marc Baumgartner, skyguide (Geneva) and President and CEO, International Federation of Air Traffic Controllers' Associations

Dr Andrew Cook, Senior Research Fellow, Department of Transport Studies, University of Westminster

Dr Nigel Dennis, Senior Research Fellow, Department of Transport Studies, University of Westminster

Ben Van Houtte, Former Head of Unit, Air Traffic Management and Airports, European Commission, Brussels

Dr Arnab Majumdar, Lloyds Register Research Fellow, Centre for Transport Studies, Imperial College London

Dr Nadine Pilon, Responsible for Prospective Studies, EUROCONTROL Experimental Centre Strategy Unit, Brétigny sur Orge

Graham Tanner, Research Fellow, Department of Transport Studies, University of Westminster

Dr Victoria Williams, EPSRC Advanced Research Fellow, Centre for Transport Studies, Imperial College London

Foreword

Air Traffic Management (ATM) aims at ensuring the safe and efficient flow of air traffic. This looks simple and evokes control towers to the layman. It is, in fact, a complex and largely unknown subject, whose many facets are covered in massive specialist documentation, but very seldom addressed in comprehensive and authoritative books. Both first-time and expert readers will find these facets very ably presented by leading specialists in this book. I must say I read it with interest and learnt!

ATM is moving from an art, whose origins are exposed vividly by Marc Baumgartner, a prominent figure in ATM, to a sophisticated industry with wide implications for airspace users, airports, travellers, shippers, and society at large. ATM is reaching such economic, social and environmental significance that it hits political circles. The creation of the Single European Sky was explicitly included in Mr Romano Prodi's presentation of the Commission's work programme in 2000. The Single European Sky regulations, whose adoption in 2004 is a landmark in the EU reform of ATM, are presented by their 'father', Ben Van Houtte.

The different facets of ATM, such as flight planning, sector capacity and cost of delays, environmental impacts and demand to be expected from future developments in air transport are examined as well, and rounded off by my colleague Nadine Pilon.

May I wish readers much pleasure and benefit in reading this book.

Xavier Fron
Head, Performance Review Unit
EUROCONTROL

Preface

This book is designed to offer the reader a single source of reference on the key subject areas of air traffic management in Europe. It brings together material that was previously unobtainable, hidden within technical documents or dispersed across disparate sources. With a broad cross-section of contributors from across the industry and academia, the book aims to offer an effective treatment of the key issues in current, and developing, European ATM. It explains the principles of air traffic management and its practical workings, bridging the academic and operational worlds to give an insight into this evolving field, with a number of fresh perspectives brought to the text. On-going research and developments are closely integrated into the themes, demonstrating the likely directions of future ATM in Europe and the challenges it will face. It is hoped that the book will appeal to both aviation academics and practitioners, equally for those whose area of expertise is outside ATM but want a clearly elucidated source of reference, as to those wishing to broaden existing knowledge. It is anticipated that many readers will already have expertise in one or more of the chapters' subject matter, but wish to develop a further understanding of the areas covered in others, taking advantage of the many thematic and operational links which have been illustrated.

Chapter 1 sets the scene for the rest of the book, establishing in some detail the fundamental principles and practices of how European airspace has evolved, and operates today. It is thus the longest chapter and some of the concluding comments have been saved for the closing section of the book. Chapter 2 then builds on this foundation, with a detailed and specialist description of the processes of flight planning and messaging, introducing the fundamental concept of capacity management, which is then explored in greater detail through numerous practical and research-based concepts in Chapter 3.

Chapter 4 addresses the inevitable consequences of a system operating under the increasing challenges of capacity constraints – delays, and the costs thereof. No book on air traffic management these days can be complete without a consideration of the concomitant environmental impacts of aviation, and the challenges these present us with; Chapter 5 explores a wide range of such issues. Chapter 6 then sets the developing story into the broader context of the future of air transport operations in Europe, which are intimately bound with ATM through issues such as fleet development, aircraft utilisation trends, airline networks and efficiency.

Chapter 7 offers an invaluable and detailed discussion of how some of the future challenges are being addressed through reform of European ATM, whilst Chapter 8 closes with an exploration of a new area of vital development in ATM – its two-way, evolving relationship with society. The book closes with a comprehensive list of further information resources and a concluding look ahead.

Acknowledgements

Air traffic management is indeed a singular affair. Some aspects remain cast in stone for decades, whilst others change more rapidly than the time it takes to create a book about them. As chapters were being religiously compiled by the contributors, the industry was furtively changing a manual here and a policy declaration there. Nevertheless, a book has been produced which, it is hoped, strikes the right balance between establishing the fundamental principles with clarity, and having many up to the minute examples to bring to the text a certain immediacy.

I would like to thank all the contributors for their painstaking investment of effort in their chapters. My first impression on reading each was that their knowledge and first-hand experience was literally jumping off the page. They made this book possible through such efforts: my thanks to them all, and apologies in equal measure for any oversights in the final version, the blame for which lies entirely at my door.

In addition, I am indebted to Gerhard Meise (Lufthansa Systems Aeronautics) and Gerhard Berz (EUROCONTROL) for invaluable suggestions regarding certain technical sections of the book, drawing on their expertise and personal experience. Sincere thanks are also extended to Xavier Fron (Performance Review Unit, EUROCONTROL) for kindly agreeing to write our foreword. Further thanks to EUROCONTROL for permission to use outputs from SkyView2, in Chapter 2.

My final word of wholehearted thanks goes to my colleague, Graham Tanner. Over the two years it has taken to produce this book (with quite a sharp peak towards the end!) he has unfailingly and tirelessly given enormous amounts of his expertise and time – time which could otherwise have been spent with his new daughter. I hope that his generous investment can now be reversed.

Chapter 1

The Organisation and Operation of European Airspace

Marc Baumgartner
IFATCA and skyguide

1.1   Introduction

In this chapter we will set the scene for much of the rest of the book. We will first explore the very beginnings of air traffic control, before showing how the situation has evolved from the use of flags, to satellites, and describing the organisation, operation and regulation of airspace in Europe.

During the first years of aviation, the density of air traffic was sufficiently low that it was possible for the captain alone to be responsible for the safety of the aircraft. It was therefore up to him to take necessary measures to avoid other aircraft, obstacles on the ground and terrain. With the spectacular increase in the number of air movements and the substantial acceleration in the development of traffic, the captain gradually lost the ability to carry out all the manoeuvres required to guarantee flight safety – and ATC was born.

In fact, little progress was made with the infant profession until after World War II, whence modern aircraft were pushing the boundaries, and testing notions of absolute sovereignty of airspace. Governments were forced to take some action to ensure safety and efficiency whilst guaranteeing the freedom of the skies. Better route structures were initiated, more efficient radios and navigation aids were introduced, and international agreements struck. The International Civil Aviation Organisation (ICAO) established procedures and promulgated standards that most national administrations subscribed to.

Before our first chapter gets underway, however, we will explain some basic terminology. Although ICAO does not use the term 'air navigation service(s)' (ANS),[1] we have chosen to use this term in order to make the following explanation clearer. This term comprises four main components:

- communication, navigation and surveillance (CNS);
- meteorological services;
- air traffic management (ATM);
- 'auxiliary' aviation services.

---

1   ICAO refers to 'Air Traffic Control Services' and 'Air Traffic Management'.

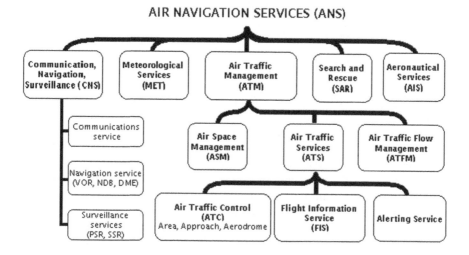

Figure 1.1   Summary of Air Navigation Services (ANS)

Source: adapted from: EUROCONTROL, 2006h.

ATM, includes all the services related to air navigation, that is:

- Air Space Management (ASM);
- Air Traffic Service(s) (ATS);
- Air Traffic Flow Management (ATFM).

The task of ASM is to plan and publish the management of airspace, divided up into air routes, civil and military control routes and areas reserved for airports, while at the same time guaranteeing the safety and fluidity of traffic. Together, *ASM* and *ATFM* support the use of the available airspace effectively, including airport capacity, by minimising waiting times (see also later comment on ATFCM).

Whilst this chapter will now develop the foundations for understanding ATC, Chapter 2 will focus on how flight planning and flow management works, with an elaboration in terms of understanding capacity, in Chapter 3. These early chapters will thus build up the picture of air traffic management, before we turn to some of its impacts, in terms of the financial costs of delay, and other implications for managing its growth in terms of the environment and society.

1.2   The Historical Context and Development of ATC

1.2.1   Early Beginnings of Flight and ATC

Air traffic control (ATC) is one of the world's youngest professions. Like many modern professions, it has developed from the humblest beginnings into a highly sophisticated and technology dependant occupation. With ATC there was no 'big bang' – it wasn't discovered or invented but has evolved gradually, driven by demand. Circumstances have dictated that it developed slightly differently from region to region, from country to country and even from city to city. The basic principles remain the same, however, whether one is using highly sophisticated synthetic radar displays and employing satellite communications, or making do with antiquated, procedural control methods – with World War II vintage high frequency radio equipment. ATC is the science (some say the art) of maintaining safety by keeping air traffic separated whilst at the same time ensuring expedition and efficiency. At times, these concepts do not lend themselves to complementary action and it is then that a controller is tested and his or her skills come to the fore.

Once man became airborne in a heavier-than-air machine, his ingenuity and developing technologies permitted him to fly higher and faster. Much the same may be said of ATC. To safeguard aviation, ATC has also employed developing technologies to manage the traffic. Unfortunately, the terrestrial developments have never kept pace with the airborne improvements and it has been in the area of this technology gap that the national air traffic control systems have sometimes been tested, often found wanting, and frequently from whence the protests and controller disputes have derived.

Within two decades of the Wright brothers changing the concept of travel, 'air traffic controllers' appeared waving flags. It is possible to theorise that Wilbur Wright was the world's first air traffic controller with Orville a close second. They did not need to file a flight plan nor seek permission to take off or land. Even by 1905, when their *Flyer No. 3* was capable of half-hour flights, it was still the only aircraft in the air.

Before long, the need for an operational watch over aircraft in flight prompted the institution of air traffic control. Firstly, it was the need to know where an individual aircraft was that led airline companies and, later, national institutions, to maintain such a watch in the event of something going wrong so that action would be speedily taken. To do this efficiently, the new invention wireless was to be utilised. Since there were precious few of those about, much cooperation was also needed. Soon after World War I ended this was available and in Europe national military and post and telecommunication authorities (who employed professional radio operators) made available their ground stations for the relay and exchange of information. Then, as more and more aircraft took to the skies, the need to keep those aircraft apart, initially as they were manoeuvring on the ground and, later in the air, became paramount.

1.2.2   The First Commercial Airlines

Aircraft development was particularly accelerated by World War I, although this put paid to much civil aviation as the emphasis switched to military applications. It was shortly after the end of World War I that people began to see that aircraft raised possibilities of profit – and commercial aviation was borne. The honour of being the world's oldest continuously operated airline goes to the Dutch carrier Koninklijke Luchtvaart Maatschappij (KLM), which was formed in 1919.

By the end of 1920, European airline development was such that KLM, two French airlines, and a Belgian airline were all flying between London and the Continent. The fare from London to Paris was £10 on the British airlines and £6 on the French carriers, the difference being brought about by government subsidies granted to the French companies. These subsidies were to prove fatal to the British carriers and by February 1921 the fledgling British airline industry had collapsed. This embarrassment prompted the British government into a subsidisation policy of their own and several airlines were resurrected.

At the same time, the carriage of wireless became commonplace and, in 1927, mandatory for aircraft carrying ten or more persons. It was this carriage of the recently invented wireless that enabled contact with, and control of, the aircraft beyond the visual confines of the airfields. Britons Alcock and Brown made the first non-stop transatlantic flight when they flew from Newfoundland to Ireland in June 1919 and the Australian brothers Ross and Keith Smith completed the first flight from the UK to Australia, in the same year. With aviators such as these proving the feasibility of long-distance flight, the commercial implications were quickly realised and airlines sprang up everywhere in the subsequent decades.

Since profit margins on the commercial flights of the day were usually very thin, it was a requirement that, as far as possible, these flights operated directly from A to B. It was when other aircraft started flying simultaneously from B to A that people began to realise that they may have a problem on their hands! However in the main, it wasn't this aspect that prompted the implementation of air traffic control. Even by the time of World War II, there still weren't too many aircraft about, at least not yet enough to warrant traffic separators. The bigger problem still was reliability and keeping track of aircraft in case something went wrong. We need to remember that these early flights were all conducted at low levels and were thus very much susceptible to the vagaries of weather and constrictions of terrain. Also, there were precious few en-route radio navigation aids to assist, so nearly all flying was conducted using visual observation.

Although there weren't large numbers of aircraft taking to the skies, administrators realised that aviation was a burgeoning industry and, in the way of administrations everywhere, decided that regulations were required and that some standardisation should be applied. This was particularly important in Europe with its multiple national boundaries and languages and, in one of the lesser known decisions emanating from the Versailles Peace Treaty, the International

Convention for Air Navigation (ICAN) was born. Nineteen nations[2] signed this convention giving weight to the development of 'General Rules for Air Traffic'.

1.2.3 The Birth of Air Routes, Controllers and Advisory Services

Apart from ICAN regulations, in a practical sense, ATC probably has its earliest roots in the London – Amsterdam/Brussels/Paris traffic growth. Following the world's first commercial mid-air collision on 7 April 1922 over France, measures were taken to ensure it wasn't repeated. These measures included the carriage of radio and organising a defined set of routes for all to follow visually. London's Croydon airport was expanding to cater for increasing traffic. The duties of the aviation staff at Croydon were to work out an aircraft's time of arrival, assist with take-offs and generally be with the pilot and arrange whatever they wanted! As is often the case, another incident prompted the further expansion of these officers' roles. A minor collision between an arrival and a departure at Croydon resulted in a Notice to Airmen (NOTAM) from the Department of Civil Aviation which told departing pilots to obtain their order of priority and to await the signal from the 'controller' to take off. This represents the first recorded authority for airport staff to make a decision on the 'control' of traffic. This signal was to be the waving of a red flag but it soon became obvious that this signal could not be seen from all over the Croydon airfield as it had a slope to one side. To remedy this, the flag waver was moved to the first floor balcony surrounding the observation hut.

In July 1922 the Croydon observation hut was rebuilt with a glass upper storey. The purpose of this, however, was not to afford the officers a degree of comfort, or even to enable them to see aeroplanes, but to test a new direction finding wireless receiver. A little later this 'tower' became the centre for all wireless communications and one of the operators also developed a flight progress display. To keep track of the aircraft, Mr Jimmy Jeffs stuck coloured pins into a map and with the aid of aircraft reports, and on his own estimates, moved them along the route. He soon added little flags to the pins detailing call signs and altitudes. If it appeared that two aircraft would pass close to each other, he was able to advise them of their proximity, which alerted the pilots to be extra vigilant. Thus was born the first 'advisory service'. NOTAM 109/1924 amended the rules for take-off permission and, in doing so stated:

> When the aircraft is visible from the control tower, permission to depart will be given from the tower

This was the first time that the term 'control tower' was used. 1927 saw the start of the compulsory use of the wireless service at Croydon by commercial operators,

---

2   Despite being present in Versailles, the United States did not sign the ICAN convention. It wasn't until 1926 that it embarked upon a programme to establish its own air traffic rules and it followed this in 1927 with a start on a 'Federal Airways System'.

the introduction of the 'Q code'[3] and the first use of Standard Departure Routes. It is interesting to note that these short-lived Standard Departure Routes were not for traffic separation purposes but were instituted as noise abatement procedures following complaints from local residents! (We will return to the theme of noise in Chapters 5 and 8.) 1928 saw the introduction of defined (and compulsory) reporting points on the London-Continent routes.

Like the Europeans, the USA introduced ATC by linking the communications service with visual aids: signalling lamps. The Federal Airways System linked the major cities and basically comprised a network of (low frequency) radio beacons enabling accurate track keeping. After a demonstration of 'instrument' flying in 1929, the USA settled upon standards to cover this type of activity and, in 1933, laid down Instrument Flight Rules (IFR) qualifications. With this progress, the need to separate aircraft flying at night and in bad weather was recognised. Formal rules to handle the traffic by 'airway traffic control centers' were subsequently developed and such centres at Newark and Chicago resulted. The world's first IFR centre was commissioned in Newark, in 1935.

Blackboards and map tables later gave way to paper flight progress strips in movable holders on flight progress boards. In 1938, when the Civil Aeronautics Act established Civil Air Regulations, pilots were *required* to comply with ATC instructions.

By 1942, around 500 controllers manned IFR centres around the United States, covering more than 40 000 miles of air routes. The final major step in this progression came in the early 1950s when adequate air-ground communications facilities were provided for the centres to enable direct controller-pilot communications, thus eliminating the hitherto laborious and time-consuming practice of relaying all instructions *via* the airline dispatchers. (We will return to IFR flight rules later in this chapter, and discuss commercial flight planning in detail in Chapter 2.)

1.2.4   The Emergence of Airways, Radar and VORs in Europe

Whilst raw display radar for terminal area use was first introduced simultaneously in Sydney and Melbourne in 1959 (with the first en-route radar centre later to be opened in Sydney, in 1965), it was The Netherlands which was at the forefront of the introduction of modern techniques to handle the burgeoning traffic in Europe (and was, incidentally, home of IFATCA's first President). Radio facilities became more widespread during World War II, but the war totally destroyed Schiphol and ATC literally had to start again from scratch. In the control room,

---

3   The 'Q code' is a standard collection of three-letter codes, which all start with the letter 'Q'. It was initially developed for commercial radiotelegraph communication, and later adopted by other radio services, such as for maritime communications, to promote radio clarity, particularly for international speakers. The code set was adopted by a convention in London, in 1912, with codes in the range QAA-QNZ reserved for aeronautical use. Although their use is quite rare these days, a few terms persist, such as 'QNH' in aviation (see later), to promote unambiguous communications.

the main control tool was a large sheet of paper, which horizontally depicted the distance from Schiphol and the vertical axis represented the time from landing. Different coloured lines were drawn from the centre to represent the air routes. Call signs, altitudes (then in metres), descent and climb arrows were written in in pencil. 1952 saw the implementation of airways in The Netherlands (the first in Europe) to replace all the direct routings which had become overcrowded and unsafe. In 1953, a surveillance radar was installed, thus enabling a considerable increase in the movement rate. In 1952, for instance, Schiphol could only handle about ten arrivals an hour under procedural control. With the new radar, this rate increased to approximately 25 per hour.

The increasing speeds and higher cruising levels of aircraft soon highlighted the limitations of early radar and a new surveillance radar[4] was developed, which provided coverage far beyond the lateral boundaries of the new airways and to a height sufficient for the emerging jet airliners.

With the raising of the upper limit of their controlled airspace from 15000 feet to 25000 feet and the earlier replacement of Non-Directional Beacons (NDBs) with VHF Omnidirectional Ranges (VORs),[5] the Dutch controllers now had the tools they needed to satisfy the fundamental demands of air traffic control.

1.2.5   *Radar Separation Develops Further – SIDs and STARs Emerge*

Schiphol controllers had moved to automatically routing traffic on the narrow airways, clear of military areas, and providing radar 'vectors' (headings) to final approach. Initially, simple procedures were prescribed – still reflecting the concept that it was a procedural system with radar back-up. Soon, however, becoming more familiar with their equipment, the controllers became involved in formulating full radar procedures and radar separation standards. Eventually, after some debate, a dual airways system was created with holding stacks at Terminal Manoeuvring Area (TMA) entry points, Standard Instrument Departures (SIDs) and standard arrival tracks. They also developed what is believed to be the world's first 'Planning and Executive Control System', a method of planning control decisions at flight progress boards but having those decisions implemented by another controller at a radar console. Not only in The Netherlands, but elsewhere in Europe, conflicts between civil and military use were evident: a theme to which we shall often return.

---

4   Not to be confused with Secondary Surveillance Radar (SSR), the first of which in Europe was not installed until 1962 in France.

5   We will discuss these 'beacons' in detail later.

1.3   An Overview of International Regulation and Coordination

This book focuses its attention on 'Europe', which we will later define to include certain countries from the former economic region of Comecon,[6] but to exclude the Russian Federation. We will now examine the international players at the organisational level of air traffic control.

1.3.1   ICAO

In the wake of studies initiated by the US and the Allied Forces, the American government invited 55 states in November 1944 to an international civil aviation conference in Chicago. Thirty-two of the states which participated in the meeting established the International Civil Aviation Organisation (ICAO). However, even more important was the creation of the 'Chicago Convention'[7] on aviation, which set the foundations for the rules and regulations concerning air navigation in all its aspects, and which enabled a common air navigation system to be created around the world.

Headquartered in Montreal, ICAO today manages all aeronautical spheres and establishes world standards. It currently has 189 member states and comprises an Assembly, a Council (33 states), auxiliary bodies and a Secretariat. The main executive agents are the Chairperson of the Council and the Secretary General. The Assembly, made up of representatives of all the contracting states, is the sovereign body of ICAO. As executive body, the Council ensures the continuity of the management of ICAO's work. It is the Council which adopts the recommended standards and practices, which are grouped in the annexes of the Chicago Convention.

ICAO is divided into ten regions managed by seven regional offices, which are responsible for the coordination and application of regional directives. The ten regions are as follows: Africa and the Indian Ocean, Asia, the Caribbean, Europe, the North Atlantic, the Middle East, North America, the Pacific and South America. The Paris regional office deals with the activities of Europe and the North Atlantic. We shall return to the role of ICAO in the context of the evolution of European airspace, and the Single European Sky initiative, in Chapter 7.

We conclude these comments on ICAO by noting that it produces a number of key reference documents for the industry, a key example being what is commonly referred to as 'Doc 4444': *Procedures for Air Navigation Services – Air Traffic Management* – see ICAO (2001).[8]

---

6   The Council for Mutual Economic Assistance ('SEV') of the former Communist Bloc.

7   Formally the 'Convention on International Civil Aviation', signed in Chicago on 7 December 1944. Through amendments and annexes, this remains a dynamic reference for the industry.

8   For consistency of referencing with other texts, we shall also cite this as 'Doc 4444', and this is cross-referenced in the bibliography of this book.

1.3.2   ATC Guilds and Unions

Notwithstanding the formation of ICAO following World War II, there remained local interpretations and national approaches to much of the problem solving in ATC. In large part, it was these uncoordinated approaches and incompatible solutions, and the ensuing problems that they created, that prompted controllers to look beyond their own borders and confer with their colleagues. So, whilst there was now in existence an International Civil Aviation Organisation which established procedures and promulgated standards that most national administrations subscribed to, it was really the individual controllers who took it upon themselves to initiate true grass roots international action to solve their problems.

Right across Europe there existed pressing demands that, to the controllers, their administrations were not adequately addressing. The controllers were in a mood to do something about it. They certainly had amongst their numbers some individuals with the professional skills and necessary acumen and now, it seemed, the climate was right for them to collectively seek change based on actual operational experience. In so doing, their common aim was to make air traffic control not just a job but a profession – and truly international.

By the 1950s, technology was starting to catch up with the traffic volumes but, as always, demand was outstripping the authorities' ability to keep pace, as traffic was now doubling around every five years and the controllers' frustrations were beginning to show. In Europe, in particular, air traffic controllers were being hampered by outmoded procedures and severe airspace limitations placed on them by the constraints of multiple national boundaries, national sentiments and the effect of the Cold War.

National trade unions existed but, in the main, they were ill equipped to cater for the needs of such a young and rapidly expanding specialist profession. Some local professional organisations had sprung up in attempts to deal with the multiple problems facing the controllers but it became apparent that, because of the universality and nature of the problems, something more was needed.

The idea of a world body for controllers was proposed in 1956 by the first Chairman of the Israel ATC Association. In November 1959, some 39 delegates from 14 countries gathered for discussions in Frankfurt, in an attempt to coordinate their approach to the problems. In September 1961, the founding meeting of the International Federation of Air Traffic Controllers' Associations (IFATCA) was held in Amsterdam. The charter and the objectives remain the same today and are just as relevant. What had started as a gathering of fourteen associations in the late 1950s is now a worldwide body representing more than 50 000 air traffic controllers, in over 130 countries.

IFATCA is an independent, non-political, and non-industrial professional association. Among the aims of the Federation is the promotion of safety, efficiency and predictability in international air navigation, along with safeguarding the interests of air traffic controllers. In order to accomplish these aims and objectives, it cooperates closely with various other aviation authorities and institutions. These include ICAO, the International Air Transport Association

(IATA), the International Federation of Air Line Pilots' Associations (IFALPA), the International Labor Organization (ILO), the International Transport Workers' Federation (ITF), and EUROCONTROL (introduced below).

1.3.3   The European Context

In the earlier days of aviation, numerous states themselves financed air navigation services (ANS) but later refused to continue doing so, and almost all EU member states have now set up corporate entities for this purpose. This new situation gave rise to the appearance of service providers who have become financially autonomous, primarily through levied user charges. Irrespective of their funding, these providers are commonly referred to as Air Navigation Services Providers (ANSPs).

According to ICAO, ANS should be provided by independent authorities, independent entities or companies established to operate these services, rather than by civil aviation authorities, as is sometimes the case today. We can describe three basic forms of ANS at national level:

a) a government department depending on the state budget with staff having the status of civil servants (this used to be common in Europe in a previously less-regulated environment);
b) autonomous bodies belonging to the public sector, which are separate from the state, while still remaining state property, i.e. with functional separation (commonplace now in Europe);
c) partly or fully privatised.

An example of the rarer (c) is the British ANSP, NATS (National Air Traffic Services). NATS is a public-private partnership. The airline group (a consortium of seven airlines) has the majority of voting rights and 41.9 per cent of the shares. The UK Secretary of State for Transport (i.e. the government) owns 48.9 per cent of the share capital, British Airports Authority 4.2 per cent and NATS employees the remaining 5 per cent. NATS is governed by a Strategic Partnership Agreement between the Secretary of State, the airline group and BAA.

Each European state is responsible for providing air traffic services as a public service and has sovereignty over the airspace above its territory. For operational purposes, a state may delegate part of its airspace to a neighbouring state. These themes will be discussed further in Chapter 7, in the context of the emergence of the Single European Sky initiative.

1.3.4   EUROCONTROL and ECAC

EUROCONTROL, the European Organisation for the Safety of Air Navigation, currently has 37 member states (including all EU member states, except for Estonia and Latvia), was established in 1960 and is currently governed by the Protocol consolidating the EUROCONTROL International Convention for the Safety of Air Navigation (see Chapter 7). It provides skills and technical expertise for aeronautics and flight across Europe and is working on the building of a homogeneous pan-European ATM system, to fully respond to the continuing increase in air traffic, while respecting the imperatives of maximum safety, cost reduction and conservation of the environment. EUROCONTROL is both a civil and a military air navigation services agency, as we will explore later.

Table 1.1   Summary ECAC data for 2006

Feature	Number
Control sectors	600
ACCs* (mostly en-route)	70
Number of flights on busiest day	31 914
Number of aircraft simultaneously airborne	over 3000
Annual flights (IFR GAT**)	9.6 million

*   Area Control Centre (to be discussed later). Known in the US as an Air Route Traffic Control Center (ARTCC), and in French as a Centre de Contrôle Régional (CCR).

**   General Air Traffic, or 'civil' aviation (to be discussed later).

EUROCONTROL's Central Flow Management Unit (CFMU) manages[9] the flows of the 42 states which participate in the European Civil Aviation Conference (ECAC) to ensure that user demand (i.e. from the airlines) does not overload the capacities offered by the infrastructure (i.e. en-route and in Terminal Manoeuvring Areas), some key features of which are quantified in Table 1.1 (derived from a number of EUROCONTROL sources). It does this through processes which will be described in detail in Chapter 2, but which essentially involve the staggering of demand in time and space, through a concept known as Air Traffic Flow Management (ATFM), which has more recently been broadened in its frame of reference to include capacity management, such that the term ATFCM (Air Traffic Flow and Capacity Management) will often be encountered. Nevertheless, the essential processes of messaging and slot allocation are still best referred to as ATFM.

Since EUROCONTROL plays a key role in European ATM, with a very wide range of programmes, initiatives and bodies which influence operations and strategic planning in almost every aspect, it will come as no surprise to the reader that EUROCONTROL will be referred to very often in this book, in a range of different contexts. In particular, current, operational contexts will be discussed in Chapters 2 through 4, and the part played by EUROCONTROL

---

9   Note that this control area is thus larger than that of EUROCONTROL's member states. The geographical scope of flow management control is described in greater detail in Figure 2.1 and its supporting text (Chapter 2), where, for example, it is apparent that Iceland is in ECAC but is only in what is referred to as the 'ATFM Adjacent Area'.

and the European Union in the development of the Single European Sky will be dealt with in Chapter 7.

1.4   The Organisation of European Airspace

Having described the development of ATC in Europe, and the authorities and bodies involved in its running, this section will explain the organisation and functioning of modern-day, European airspace. We will move from the over-arching concepts of national boundaries, sectorisation and military areas, through navigation aids and airways, to the specifics of the TMA, SIDs, and STARs.

In Annex 11 of the Chicago Convention (ICAO, 1998) the international regulations concerning air traffic services (ATS) applied in almost uniform fashion throughout the world are described. A distinction is made between three types of service:

- air traffic control services;
- flight information services;
- alert services (e.g. to warn the relevant bodies when an aircraft in difficulty is in need of assistance from the search and rescue bodies, and to give these bodies the necessary cooperation).

Air traffic control services deal with the movement of traffic in the airspace controlled. Three main organisational distinctions may be made (Skyguide, 2001):

- Whereas commercial and private ('civil') flights are referred to as General Air Traffic (GAT), military training flights and operations are known as Operational Air Traffic (OAT). The vast majority of civil use is constituted by commercial airline movements. A pilot planning a flight can choose between two different types of flight rules:
- Visual Flight Rules (VFR) means the pilot is subject to certain meteorological constraints (in particular, minimum visibility criteria). The pilot is responsible for maintaining visual contact with other airspace users and for determining his route with the help of geographical landmarks.
- Instrument Flight Rules (IFR) means the pilot can begin the flight regardless of the meteorological criteria. The instruments are navigation aids (radio beacons, instrument-assisted landing systems and satellites – see later). On-board navigation systems and indications transmitted by air traffic controllers on the basis of radar data make up for the lack of visibility and provide all the information required for the pilot to be aware of his position at all times. We will look at commercial IFR flight planning in detail in Chapter 2.

Table 1.2   Air traffic control services

Abbreviation	Appx. number in Europe (ECAC)	Type of control	Remarks on control
TWR (Tower Control Unit)	400	Aerodrome control (tower)	Exerted over the runways, taxiways, aprons and stands of the airport; takes charge of runway management and assistance for the take off and landing of aircraft
APP (Approach Centre/ Control)	200	Approach (or 'terminal') control	Takes place in approach control rooms and deals with the approach and/or departure phases on the runways, and until aircraft are airborne and at least clear of the aerodrome; note the use of 'TRACON' in the US: Terminal Radar Approach Control
ACC (Area Control Centre)	70	Mostly en-route	Mostly concerns the 'main' flight phase and is carried out in air navigation en-route centres

1.4.1   Flight levels and QNH

In air navigation, altitude is measured as the vertical distance from Mean Sea Level (MSL). Pressure altimeters in the cockpit display the 'indicated altitude', which is shown in feet Above Mean Sea Level (AMSL). The default height measurement is calculated assuming a standard pressure of 1013.25 mb (hPa) at MSL, using the atmospheric pressure outside the aircraft.

However, this needs to be corrected for the *actual* pressure at MSL, at that particular time and location – known as 'QNH' (from the Q-code radio communication system – see footnote 3). QNH is entered by the pilot using a subscale-setting knob or dial, and the QNH setting in use is shown in a small window in the altimeter scale. The altimeter looks rather like a watch, with the hands showing the altitude in feet, and the window where the day and date might be shown in a watch, showing the QNH setting. Without a correct setting of the subscale (QNH), the reading of the altimeter is meaningless. Accurately set, altitudes should be displayed to within at least ± 50 ft.

The flight levels (FLs) at which aircraft operate are also known as 'pressure altitudes'. A pressure altitude of 5500 ft (i.e. 5500 ft higher than the standard, 1013 mb pressure level) is typically written as FL 55 (the last two zeros being omitted). Flight levels are usually separated by at least 500 ft, e.g. as a sequence: FL 45, FL 50, FL 55. A flight level is thus correctly defined as a surface of constant atmospheric pressure.

Particularly at lower levels and in respect of terrain clearance, problems could arise when QNH changes, or when flying into an area of higher or lower pressure than your current location. To avoid such potential problems (and those of traffic separation which could arise if aircraft approaching each other from different areas still had their origin QNH settings on their altimeter subscales) aircraft use a regional pressure setting, or regional QNH, also known as Area QNH. Its value is updated to pilots, by ATC, at least every hour and this is based on the lowest forecast QNH value for the next hour in the whole region to which the setting applies. This means that any small, systematic error puts pilots at a slightly higher altitude than that indicated (since QNH is at sea level, or slightly higher). Since a given flight level will be a different number of feet off the ground depending on the prevailing local pressure, this has practical manifestations. For example, the minimum stack level of the Heathrow in-bound holds can either be FL 70 or FL 80, depending on QNH.

An update on QNH might be communicated to an aircraft even by ground control during a taxi-in, i.e. as soon as the value is revised, such that all aircraft at an airport have the same revised setting. The UK CAA's *Manual of Air Traffic Services – Part 1* ('CAP 493' – CAA, 2006c), states that when the pilot of an aircraft requests start-up or taxi clearance the following information shall be given:

runway in use;
surface wind direction and speed, including significant variations;
aerodrome QNH;
outside air temperature (turbine-engine aircraft only);
significant meteorological conditions (e.g. RVR, marked temperature inversion).

Those items which are known to have been received by the pilot may be omitted. This might include data already passed during taxi-in, or through 'special' aerodrome meteorological reports. CAP 493 states that 'specific improvements and deteriorations of any of the items in a routine report are supplied in a special report'. These are issued between routine reports and contain only those items which are affected. One of the criteria for raising special reports is if QNH changes by '1.0 millibar or more'.

Approach controllers also need to keep a close watch on the (real-time) QNH reading at their station, in order to communicate the correct value to aircraft entering their control area. If a controller communicated a value of QNH 1030 mb, then noticed it had changed to 1032 mb since they last looked, they would re-contact the aircraft with an update, even though such small changes only represent an error of some 60 ft. It is noteworthy, however, that CAP 493 seeks to avoid *unnecessary* radiotelephony (RT):

Air traffic service units are to have available the Regional Pressure Setting for the altimeter setting region in which they are situated and appropriate adjacent regions. These values are to be passed to pilots when requested or at the discretion of the controller. However, a pressure setting shall not be volunteered if a controller is uncertain that it is appropriate to the flight.

On entering a TMA, a QNH setting for the (major) aerodrome situated in the TMA may be required. Light aircraft may also be required to set QNH for such a TMA if they are flying *beneath* it (e.g. further out from the aerodrome, where the TMA does not extend to ground level) to ensure they do not inadvertently penetrate the controlled airspace from below. Such practices vary from state to state.

1.4.2   The Transition Altitude and Transition Layer

When cruising above a certain height, known as the Transition Altitude, a common ('assumed') QNH setting of 1013 mb is used. The transition altitude used depends in part on terrain clearance issues but saves pilots continually re-setting the subscale. The transition altitude may even vary between sectors in the same state. In the UK, the highest terrain (Ben Nevis) is at 4406 ft AMSL, and the transition altitude is generally at 3000 ft AMSL, but with several variations (e.g. in some areas it is 6000 ft) depending on terrain and traffic densities. Plans to standardise the transition altitude across the UK in 2005 (at 6000 ft) were not realised, and this still varies both inside and outside controlled airspace. In the US, where terrain is much higher, there is a common transition altitude of 18 000 ft (FL 180) across the entire country, in contrast to the current variations which persist in Europe.

In controlled airspace, and flying above the transition altitude, aircraft operate at 'flight levels' (with 1013 mb set); at or below the transition altitude, they operate at 'altitudes' (with QNH set). For example, if ATC instructs a descent from above the transition altitude, to below it, the altitude to which the aircraft is instructed to descend is always given in feet, and the pilot informed of the QNH setting. Such a descent involves passing through the transition layer. According to CAP 493, for arriving aircraft:

Aircraft at or below the transition level are to be given the aerodrome QNH.

When an aircraft is cleared to descend from a flight level to an altitude the appropriate QNH shall be included in the same transmission. If flight level vacating reports are required the request should be included with the descent clearance. Transition level is passed to aircraft only if requested.

After QNH is assumed to have been set by an aircraft all reference to vertical position shall be in terms of altitude until the aircraft commences final approach. Vacating reports which have not been requested at the time of the descent clearance may be in terms of altitude, particularly if the aircraft has only one altimeter.

The transition layer is defined as the airspace between the transition altitude and the transition level. The transition level varies with QNH, and is the lowest flight level available for use above the Transition Altitude (which is fixed). For example, with a transition altitude of 4000 ft AMSL, FL 45 (the lowest flight level above it) would be less than 500 ft above the transition altitude if QNH were low (e.g. 1005 mb). The transition level would then be assigned as FL 50. The thickness of the transition layer thus varies with QNH.

The minimum sector altitude (MSA) is the lowest altitude which may be used in an emergency, which still provides a clearance of at least 1000 ft above all obstacles.

1.4.3   National Boundaries, Sectorisation, ACCs

The Chicago Convention stipulates in Article One that the contracting states recognise that every state has complete and exclusive sovereignty over the airspace above its territory. In Article Two, this is defined thus: 'for the purposes of this Convention the territory of a State shall be deemed to be the land areas and territorial waters thereto under the sovereignty, suzerainty, protection or mandate of such State'.

This has led to Flight Information Regions (FIRs) being established (normally)[10] according to national borders with several limitations dictated by military and ATC infrastructural needs. In rare cases, FIRs were established in a trans-national spirit, mainly due to technical and/or operational requirements. Note that although the word 'information' is used for these regions – a remnant of the earlier function of ATC – these regions do in fact define areas of *control*.

Further, a vertical split exists in some countries, where an Upper Information Region (UIR) was created. A state is free to choose the split between the FIR and UIR, since there is no harmonisation – example splits occur at FLs 195, 245 and 285. One of the recent, main improvements in the European Civil Aviation Area (ECAC) was the introduction of Reduced Vertical Separation Minima (RVSM: above FL 285, 1000 ft instead of 2000 ft) which could be used as a federating argument when it comes to the establishment of a harmonised standard for the lower level of UIRs. Currently, the Single European Sky initiative has established a mandate to create a European UIR (see Chapter 7).

However, as outlined in the ECAC airspace strategy, it is not necessarily the vertical split which is of importance to the users and the airspace managers, but rather the fact that each of the European states has adopted different ICAO airspace classifications. ICAO establishes seven airspace classifications which have different rules and obligations for the provision of air traffic control. These are denoted by the letters A through G (e.g. 'Class A Airspace') – Class A being under the strictest control, Class G under no ATC control at all. The ECAC airspace strategy is aiming at harmonisation towards just three airspace categories, and with a common transition altitude throughout Europe.

It has been mentioned several times over recent years that the European ATM system is not as efficient as similar areas of the US system and that sector

---

   10   Sometimes, in complex cases, other pragmatic solutions are found (e.g. between Switzerland and Germany where the Allied Forces decided on the FIR border, after the end of World War II).

productivity is much higher in the US. This has, of course, led to some debate in Europe. It is important to understand that sector definition (the smallest area of airspace under specific control) will be established based on various parameters: airspace classification, separation minima, technical equipment, military requirements, training and controller availability. Where controller-pilot communication is achieved by radiotelephony (RT), the limiting factor in dense continental airspace will be the number of aircraft an operator can handle at any given moment. Based on this empirical figure, a capacity for the sector is established (an issue discussed in more detail in Chapter 3). Ultimately, one controller will be able to handle only a limited number of aircraft at any given time, the maximum number depending on the size and type of sector, but rarely exceeding thirty aircraft, in a European sector. The declared capacity plan (aircraft per hour) is published to the users and CFMU. Currently, around 70 (mostly en-route) Area Control Centres (ACCs) exist in the ECAC Area, they control both upper and lower airspace and have various sizes and shapes.

1.4.4   Military Areas and Flexible Use of Airspace

There are two basic themes which will be explored in this book regarding the integration of civil and military airspace use. Whilst we will very briefly set the scene here, the operational implications for today's flight planning will be explored in Chapter 2. In Chapter 7, we will revisit the theme of the historical development of the joint demands on the use of European airspace, framing this in the context of future integration plans, under the Single European Sky initiative.

Whereas military aviation has sharply decreased in Europe since the dissolution of the Warsaw Pact in 1991, the volatility of military air traffic (e.g. the spectacular increase in a very short period during the Kosovo Crisis) has increased, and, under increased commercial pressure, much formerly 'reserved' military airspace has now became available to commercial airline use.

Responding to a need to better accommodate the changing needs of the military, and markedly increased civil congestion, the Flexible Use of Airspace (FUA) concept was introduced, and continues to be developed, under the leadership of EUROCONTROL (which, as we have noted, is a joint military and civil organisation). We will explore the practical operation of FUA, in Chapter 2.

1.4.5   Navaids and RNAV

This section discusses navigational aids (referred to as 'navaids' – which may be ground-based, e.g. radio beacons, or satellite-based) and area navigation (known as 'RNAV'). RNAV is a method of navigation based on the use of coordinates – it reduces the dependency on the location of beacons and permits aircraft operation on any desired flight path and thus allows ANSPs to offer more cost-effective solutions to airspace users, by making more efficient use of the available airspace, thus bringing capacity benefits. This positioning of aircraft without reference to beacons was an important breakthrough.

Basic-RNAV (B-RNAV) means having a navigation performance equal to or better than a track-keeping accuracy of ±5 NM for 95 per cent of flight time. Precision-RNAV (P-RNAV) requires ±1 NM for 95 per cent of flight time (although many of the systems in use today are already far better than this). Although it may have been acceptable in the days when there was only B-RNAV to refer to it simply as 'RNAV', it is now considered poor practice in specific, technical contexts to drop the 'B' or 'P' designator.

ATC procedures (such as a Standard Instrument Arrival – see later) describe what the pilot and aircraft should do at various geographical positions. Such procedures are described, for example, in the state's AIP (Aeronautical Information Publication – often, arguably unhelpfully, referred to as the 'Air Pilot'). A 'fix' (determination of the aircraft's position) may be associated with a change in permitted speed or altitude, with a change of which controller should control the aircraft, or with an action to be taken by the pilot. An example is an Initial Approach Fix, which usually signifies the start of the landing phase known as 'initial approach'. This type of fix is similar to a 'reporting point', which is defined by ICAO (Doc 4444) as: 'A specified geographical location in relation to which the position of an aircraft can be reported' (note use of the word 'can' – see later). For example, an Initial Approach Fix may also be a reporting point at the end of a STAR, with an instruction in the AIP such as: 'Clearance limit is normally the IAF'.

In en-route airspace, the main routes for aircraft currently consist of airways (usually with widths of 5 NM either side of the centre-line) and Upper Air Routes (with no defined width as they are always in controlled airspace). These follow straight lines between 'significant points',[11] also widely referred to as 'waypoints'.[12] If a waypoint is represented by a five-letter identifier (e.g. KONAN, FERDI, BUPAL) then it has no associated ground-based navaid – it is just a geographical coordinate. If a waypoint has a three-letter identifier (e.g. DVR, KOK, SPI) then there is an associated ground-based navaid (e.g. DME). Most five-letter waypoints are found on Upper Air Routes, although they are also found in lower airspace, and not all three-letter waypoints are shown on high-level charts.

Clearances are permissions to proceed under specified conditions and/or to a specified point, such as an aerodrome, a 'fix', a waypoint or a controlled or advisory airspace boundary. Clearances may be issued tactically (e.g. 'cleared to [navaid] overhead') and/or specified in the AIP (e.g. 'do not proceed beyond [waypoint] without ATC clearance'), for example.

---

11   A 'significant point' is defined by ICAO (Doc 4444) as: 'A specified geographical location used in defining an ATS route or the flight path of an aircraft and for other navigation and ATS purposes.' The definition used by the UK CAA (CAA, 2006c – 'CAP 493') is identical, itself citing ICAO.

12   A 'waypoint' is defined by ICAO (Doc 4444) as: 'A specified geographical location used to define an area navigation route or the flight path of an aircraft employing area navigation'. CAP 493 (CAA, 2006c) refers to waypoints, but does not include this item in the definitions of the glossary in Part 1.

A waypoint may also be a reporting point, although it is to be noted that even the presence of a 'compulsory reporting point' does not necessarily indicate that the pilot has to report the aircraft's position there. In most of Europe, the need to report position is becoming increasingly rare, partly due to improved technologies (giving ATC better positional information) and partly as an effort to reduce controller RT workload. As ICAO (ibid.) states:

> Under conditions specified by the appropriate ATS authority, flights may be exempted from the requirement to make position reports at each designated compulsory reporting point or interval ... This is intended to apply in cases where adequate flight progress data are available from other sources, e.g. radar ... and in other circumstances where the omission of routine reports from selected flights is found to be acceptable.

Such conditions could be specified in the AIP, or tactically by the controller, by such instructions as:

'Omit position reports'
'Omit position reports this frequency'
'Reports required only at [waypoints]'

c.f.   'Report passing [waypoint]'
        'Report [waypoint] inbound'
        'Report [waypoint] outbound'

According to ICAO (Doc 4444), a waypoint may be one of two types:

- Fly-by waypoint: A waypoint which requires turn anticipation to allow tangential interception of the next segment of a route or procedure
- Flyover waypoint: A waypoint at which a turn is initiated in order to join the next segment of a route or procedure.

In addition, 'company' (airline) waypoints may be inserted into the Flight Management System (FMS) of an aircraft, for example to help better navigate a fuel-saving Continuous Descent Approach, as a temporary routing to avoid weather (with ATC permission), or for use on Conditional Routes (see Chapter 2). Such waypoints may be created *ad hoc* by the pilot or be pre-programmed, and do not normally appear on official charts, or in AIPs, but rather are used as a supplement to existing waypoints.

Note: Different texts may variously use one or more of the terms 'waypoint', 'significant point' and 'reporting point' interchangeably. Of the three terms, the ICAO definition of the first two shows that their meanings are indeed very close. 'Reporting point' is the most differentiated of the three by definition, although often very similar in practice (especially considering the diminishing need for position reports). 'Waypoint' is the more generic term, and will be used henceforth in this book, notably in Chapter 2.

Returning to our discussion of beacons, and as we have seen already, in the early days air navigation was carried out under visual conditions (known today as flying according to Visual Flight Rules, or VFR). This was often assisted by flashing ground 'beacons', which were a great aid for night-time navigation in clear weather. In overcast conditions, or during the day, when they weren't turned on, they were not of great use. In order to overcome the operational limitations imposed by bad weather, electronic radio navigation systems were developed. The first such system that was globally standardised was the Non-Directional[13] Beacon (NDB). Any aircraft equipped with an appropriate antenna (typically a cable strung between the top of the tail fin and the main fuselage) and receiver could then 'home in' on NDB ground transmitters by pointing the aircraft in the direction of maximum signal strength.

Thus it became possible to fly in all weather and night conditions (if you dared) based on such ground infrastructure and the corresponding aircraft equipment. As operations relying on radio navigation aids were further developed, they became known as Instrument Flight Rule (IFR) operations. Indeed, this NDB infrastructure on the ground (in some places even with beacons every 20–30 miles along the route) was soon to be complemented by the VHF Omnidirectional Range[14] (VOR) navigation system. The VOR was probably the most significant aviation invention other than the jet engine. This is because it overcame NDB limitations (such as relatively low levels of precision – they are not suited to an RNAV environment) by providing continuous radial ranges to the station, enabling cross-wind correction and a much more accurate position determination (using either stopwatch timing or a second VOR radial).

The widespread introduction of VORs began in the early 1950s, and was itself further enhanced by later coupling with Distance Measuring Equipment[15] (DME). Since DME provides the range to the VOR in nautical miles, the position of an aircraft could now be easily transposed onto a map.[16]

All these ground-based navaids have a limited range, and provide a means to aircraft for determining their position with respect to flight paths along fixed

---

13   They are called 'non-directional' because the true course to the station could not be determined by homing, and consequently, the pilot was unable to compensate for wind drift. Early airborne NDB receiving equipment used manual direction finding, where a rotatable, directional antenna was placed in a fuselage mounted radome.

14   The basic principle of operation of the VOR is simple: the VOR transmits two signals at the same time. One signal radiates identical phase in all directions, while the other is phase shifted horizontally around the station. The airborne equipment receives both signals, looks (electronically) at the phase difference between the two signals, and interprets the result as a radial from the station.

15   Paired pulses are sent out from the aircraft and are received by the DME ground transponder. The transponder then transmits paired pulses back to the aircraft at the same pulse spacing but on a different frequency. The time required for the round trip of this signal exchange is measured in the aircraft's DME unit and is translated into nautical miles from the aircraft to the beacon.

16   Known as 'rho-theta' or 'polar' navigation.

routes. This form of positioning is always geo-referenced, i.e. the position is established by knowing the location of the navigation aid on a map.

Since the 1990s, however, Global Navigation Satellite System (GNSS) has become available. In contrast, this establishes position with respect to a coordinate frame (latitude, longitude, height), and thus supports RNAV. GNSS is already a primary means of navigation in many areas. Currently, Global Positioning System (GPS) is the only fully operational satellite navigation system to provide global and highly accurate positioning without significant coverage restrictions. In order to meet aviation requirements for integrity under IFR operations and to qualify as GNSS, basic GPS is combined (for example) with advanced airborne receiver algorithms such as Receiver Autonomous Integrity Monitoring (RAIM). It may also be accomplished by using 'conventional' ground-based navaids (such as DME) feeding an RNAV computer.

Larger aircraft may have independent, self-contained systems.[17] These inertial navigation systems are known are 'INS's or 'IRS's and extend RNAV coverage beyond that of ground-based navaids, since they are completely independent of them. Although they are expensive and their initial alignment on aircraft start-up can take some time, the time needed is usually shorter than that which the pilots require to get ready to taxi, and subsequent position updating is automatic.

Thus, contrary to some misconception, RNAV is not solely dependent on satellite navigation systems. In addition to using self-contained methods such as inertial navigation systems, it may also be accomplished by using 'conventional' ground-based navaids (such as DME) feeding an RNAV computer, or, indeed, by a combination of methods.

Aircraft can only perform IFR operations if they have the appropriate on-board instruments, according to the Minimum Equipment List (MEL) required by either the regulatory authorities or by additional airline procedures. The MEL is starting to become quite a complex issue, when we factor in RNAV systems, as they can in some cases be used to substitute conventional sensors or be a function of what equipment is needed for the intended flight plan. (See also the discussion in Chapter 2 on item ten ('equipment') of flight plans, where 'P' indicates P-RNAV capability).

Turning to TMA operations, we note that conventional navaids (such as NDBs, VORs and DME) provide service for these, as well as for en-route navigation. Final approach operations are also supported by these navaids for so called 'non-precision' approaches. They are called 'non-precision' because they do not contain vertical guidance. Precision approaches include vertical guidance and are today supported by the Instrument Landing System (ILS).

The Instrument Landing System (ILS) is a navaid which constitutes the most common precision approach currently in use and is designed to provide an approach path for exact alignment and descent of an aircraft on final approach to

---

17   Although self-contained systems, they require external navigational updates. GPS or DME/DME are most of the time coupled with the Air Data System (which comprises sensed air data, such as pressure, altitude and vertical speed, processed by the Air Data Computers (ADCs), and feeds various flight instruments).

the runway. The ground equipment consists of two highly directional transmitting systems and, along the approach path, up to three 'marker' beacons, giving distance to threshold information. In many states, markers are being replaced by DME facilities that are coupled to the ILS.

The directional transmitters are known as the 'localiser' (which indicates the centre-line of the runway) and the 'glide path (transmitter)' – a beam usually angled at 3°, indicating the correct angle of approach. Each transmits on a pre-determined channel, with 40 ILS UHF frequencies available for localisers, and another forty for the glide path transmitters. The system can be likened to a radio beam toboggan, which aircraft join typically ten miles out from touchdown, except that aircraft slow down as they descend, rather than speed up!

Above FL 95, B-RNAV is mandatory in all ECAC airspace. This 'en-route' mandate led to increasing pressure to introduce B-RNAV in TMAs also, mostly so that it could 'join up' with the en-route network. However, B-RNAV is considered insufficient for use in the more congested TMAs, particularly for arrivals and departures (SIDs and STARs – see next section) and is not intended for use below the Minimum Sector Altitude (MSA – as already defined). Consequently, there is a move to introduce standardised rules (to replace national variations) across the ECAC states for the introduction of P-RNAV in TMAs. This will improve TMA capacity (less space being required for given movements), reduce costs and RT workload, give more accurate distances to go, and improve environmental impacts such as fuel burn and noise footprints: the use of different SIDs (including Noise Preferential Routes) and STARs for daytime and night-time operations are facilitated. P-RNAV is not intended for use beyond the final approach waypoint or fix (and is thus not used for Final or Missed Approach segments). Distinct from P-RNAV, the development of Ground Based Augmentation Systems (GBAS) is ongoing, which will eventually support final approach.[18]

P-RNAV approval includes navigation data integrity and flight crew procedure requirements. The track-keeping requirements can normally be achieved using multiple DME or GPS (while some very limited use of VORs is possible, the industry is trying to get away from this: VOR is not a very good RNAV sensor, and expensive to operate). For short periods, it may be maintained using INSs, for example when taking off from a runway without appropriate DME coverage.

No ECAC-wide mandate for P-RNAV in terminal airspace is currently foreseen,[19] and non-RNAV procedures may continue in TMAs where the use of P-RNAV is not deemed necessary (and, indeed, as an alternative, e.g. in cases of loss of P-RNAV capability). Many states may, however, require P-RNAV procedures in major TMAs and for segments below the Minimum Sector Altitude or Minimum Vectoring Altitude.

---

18   Category I precision approaches first, then II/III later (referred to as 'CAT I', 'CAT II' and 'CAT III').

19   Although some airports have introduced RNAV procedures in the terminal area, these implementations have not used harmonised criteria. The purpose of P-RNAV introduction is to eliminate such local variations by making harmonised criteria available and requiring states to apply them for their RNAV implementation in TMAs.

1.4.6   Future Satellite-based Navigation and 4D RNAV

Currently, the conventional navaid infrastructure is being optimised to better meet the needs of RNAV, but this is not the way of the future. As the implementation of procedures based on a fully developed GNSS advances further, many ground-based navaids will eventually be phased out as they no longer meet operational requirements. EUROCONTROL's *Navigation Strategy for ECAC* (EUROCONTROL, 1999b), consistent with the *ATM 2000+ Strategy* (EUROCONTROL, 1998), states that in most parts of ECAC there is coverage with high redundancy provided by VORs and DME for en-route operations. Under its description of 'Strategic Steps – 2005–2010' and 'Rationalisation of ground based infrastructure', it identifies:

- NDB (withdrawn before 2010);
- VOR (withdrawn by 2010);
- DME (comprehensive coverage[20]).

Already in Europe, most aircraft fly GPS-based RNAV and overlay the VOR routes. The route structure has, in any case, already moved away somewhat from being tied to VORs. Both ICAO and EUROCONTROL support an eventual transition to GNSS, to support RNAV everywhere, including over the oceans and in remote areas.

At the same time that GPS is being modernised, the European Union is a developing a complimentary positioning system – Galileo. During what could be quite a long implementation period, ground-based back-up systems are likely to be required. Driving this development is the fact that GNSS avionics are relatively inexpensive, so all levels of users can participate in RNAV operations (both basic and precision). GLONASS (the Russian Federation's GPS analogue) now also seems to be a viable system.

The Required Navigation Performance (RNP) is a set of standards endorsed by ICAO, succinctly described in Annexe 2 of the *Navigation Strategy for ECAC* (ibid.), which comprises:

> … a statement of the aircraft navigation performance defined in terms of accuracy, integrity, availability and continuity of service necessary for operations within a defined airspace, without requiring specific navigation equipment.

> For en-route purposes currently four RNP "Types" have been defined (RNP1, RNP4/5, RNP12.6/10, RNP20), where the type number indicates the containment value in miles. The containment value is the distance from the intended position within which flights would be found for at least 95% of the total flying time.

---

20   'The DME infrastructure will continue until at least 2015, and will support RNP-1 RNAV operations adequately. Multi DME-based RNAV systems, INS/IRS with update, GNSS systems will provide the required performance.'

States must ensure that the navigation infrastructure provided supports adequately the prescribed RNP type in a specific area or on a specific route. RNP is only one parameter in the determination of separation standards.

Once the VOR and NDB infrastructure starts to be decommissioned, operations on RNP-5 and RNP-1 routes will require conformance to the requirements for RNP-5 RNAV and RNP-1 RNAV (Minimum Aviation System Performance Specification – MASPS) equipment, respectively. The situation is summarised in Table 1.3.

Table 1.3   Track-keeping performance against RNAV applicability

Track-keeping performance	Pre-RNAV MASPS applicability	Post-RNAV MASPS applicability
5 NM, 95%	B-RNAV	RNP-5 RNAV
1 NM, 95%	P-RNAV	RNP-1 RNAV
<1 NM, 95%	–	RNP-n RNAV (n<1)

Source: adapted from *Navigation Strategy for ECAC* (ibid.).

From the capacity perspective, as we will see in Chapter 3, two key steps forward anticipated for the management of en-route airspace are the introduction of real-time 4D navigation (4D RNAV – i.e. additionally including accurate time control) and the increasing delegation of separation responsibilities to flight crew, leading eventually to 'free flight' airspace. By extending the capability of RNAV, 4D-capable aircraft systems certified against 4D RNAV MASPS could support such advanced operations, and we will refer to some existing 4D capabilities in Chapter 4.

1.4.7   SIDs/STARs

To increase the capacity and flexibility of airport operations, Standard Instrument Departure routes (SIDs) and Standard Terminal Arrival Routes (STARs) have been introduced. Both are similar in many respects, and offer the pilot a pre-planned IFR procedure.

Both may be included in one type of flight plan ('FPLs'), but not in 'advance batch' flight plans ('RPLs') – see Chapter 2 for a full explanation of this. Even after the flight plan has been filed, the SID or STAR to be used might be changed tactically by ATC, depending on traffic volumes, runway(s) in use, pilot requests, weather etc. Departure and arrival clearances issued to aircraft normally confirm the SID or STAR to be used, respectively.

If a change needs to be made to a SID, before the aircraft takes off, this may be transmitted by the ground controller, or by clearance delivery, by means of radio or a datalink (Controller-Pilot Data Link Communications – CPDLC) and may be entered by the pilot into the Flight Management System (FMS). SIDs promote highly specific departure procedures (in terms of track-keeping), which is very important nowadays with noise abatement and environmental protection being major societal factors around airports (see also Chapter 8).

STARs are designed to expedite ATC arrival procedures and to facilitate the transition between en-route and instrument approach segments. Each STAR procedure is presented as a separate chart and may serve a single airport, or more than one. If the STAR needs to be changed, the pilot is usually informed of the new STAR to be used sometime before Top of Descent, whereupon it is entered into the FMS. STARs are used to streamline approach flows and to give a more regular approach to an airport. Typically, the controller will then use radar vectors (a series of headings) to line up the aircraft on the ILS, after leaving the STAR.

1.4.8   TMAs and Airports

We have already alluded to the TMA. This is a block of airspace above an airport, called a Terminal Control Area or Terminal Manoeuvring Area, but abbreviated in either case as 'TMA'. The TMA, as distinct from en-route airspace, is essentially a special type of airspace designed to handle aircraft arriving and departing the airport(s) contained within it. It excludes the aprons, although SIDs and STARs are part of it. TMAs are the most complex types of airspace, and normally only IFR flights are allowed in them. Different TMAs have different shapes. The London TMA (Class A Airspace) covers all five London airports, and extends down as low as 2500 ft close to these major airports (with a higher base level further out), and extends up to Upper Airspace (the UIR), at 24 500 ft. The London TMA, unlike many others, also contains holding 'stacks', where aircraft may be asked to circle for some moments, until there is a slot for them to land at one of the world's busiest runways.

Airports are very important players in air traffic control. The control tower, which is often the most visible architectural element of an airport, is rarely managed by the same authorities as the rest of the airport. In many of the larger airports, ATC is managed by the ANSP, and in some areas of the world, ATC follows the instructions of the airport handlers (e.g. in the US, where the dispatcher of a main hub actually influences the arrival sequence according to his requirements). Taking London Heathrow as an example, it is owned by *BAA* and the air traffic control service provider is *NATS*. Most of the ATC services are thus provided by NATS, but *stand allocation costs are* controlled by BAA.

The management and operation of a large majority of the bigger European airports are either fully, or at least partially, in the private sector (and other privatisations are planned, for example: Paris, Amsterdam and Milan). It is also to be noted that many European airports generate most of their income through non-aeronautical revenues, i.e. not through ATC.

1.5   How Controllers Handle Aircraft

In this section, we will discuss how controllers handle aircraft, including a look at
their communications with pilots and adjacent ATC sectors. Although controllers
are working within very stringent rules, it is often not appreciated just how much
flexibility and decision-making lies with them. When dealing with commercial
flights, controllers must keep the aircraft within controlled airspace (which by
definition is airspace under the direct control of ATC), but they can use the
entirety of such airspace to maintain their number one priority: safety, which, in
the main, means appropriate separation of aircraft (Michel, 1995):

> One should be aware that air controllers perform a very particular, and to a certain
> extent, paradoxical job. On the one hand, they are bound by restrictive procedures and
> a great number of very strict prescripts. On the other, they are constantly faced with
> unusual situations that require a very high degree of intellectual flexibility.

1.5.1   Sectors and Control Strips

Whether physically located in the ACC or not, each ACC has its own Flow
Management Unit/Position (FMU/FMP) which will establish a 'D-2 plan'[21] (i.e.
two days prior to the tactical day of operations), based on expected traffic and
known constraints. As the situation develops, and new information is received
and assessed, a 'D-1' plan is published and announced. During the actual day of
operation, the FMU/FMP (and possibly the airports) will adapt the plan to the
actual, tactical traffic flow, in the manner described in Chapter 2. Resultant delays,
their costs, and the proportion of overall delays caused by flow management, are
discussed in Chapter 4.

The capacity of an *ATS* unit is a function of the individual sector capacities,
and the flexibility and adaptability of the airspace configuration and sectorisation.
It also depends on a variety of operational and managerial parameters, such as
the number of available staff, tactical configuration management, and a number
of uncontrollable exogenous factors (such as the weather). Depending on their
size, ACCs are split into sectors – from two to 25 across ECAC. Each of these
sectors can handle a certain number of aircraft during a given period of time,
depending on the actual traffic pattern, the traffic complexity, and hence the
workload induced by the flights handled (the central theme of Chapter 3).

Each control sector is staffed by its own control team during most of the day
(this could be simplified during low traffic volumes, e.g. at night). With aircraft
typically crossing several sectors in one ACC, the capacity of an ACC is less than
the sum of its individual control sectors.

ACC supervisors can be planners for sectors, or for whole centres – this
depends very much on the ACC history, existing agreements, and local operational
culture. The supervisor's role can be very varied: from assigning controllers to

---

21   Which may be referred to as 'J-2' in Francophone states, where 'J' stands for
'Jour', etc.

their working positions, supervising the FMP and being the first interface with adjacent sectors, to being the representative of management and ensuring that management rules are followed.

Traffic entering and exiting the control sector is managed in cooperation with adjacent internal and/or external sectors. A given sector will be responsible for respecting the instructions and indications of the adjacent sector(s) for delivering aircraft 'cleanly', which means sufficiently separated and free of any potential conflict. Flight integration systems may differ from one centre to another, although the basic functionalities and principles are similar. We will use examples from Geneva ACC Lower Airspace (up to FL245) and the London ACC (operated by NATS, at Swanwick).

Controllers become aware of any new aircraft that enters, or is going to enter, their sector by means of a card or paper 'strip' (more correctly a 'flight progress' or 'control' strip). These strips are printed automatically some minutes (e.g. seven minutes in Geneva ACC) before the aircraft enters the sector, this being triggered by a computerised data exchange from the adjacent centre. This means that when a control strip arrives, the aircraft is not necessarily already visible on the controller's radar screen. Different centres have different designs of strip, but data are recorded systemically in pre-defined areas on the strip, and strips (and/or their holders – see Figure 1.4) are typically different colours to represent different kinds of flight (e.g. departing, arriving, transiting).

SWR665	310	350			■
5101 / SSR 2					KOR
MD80 440				MOL	
LEPA LSZH		◁ 200	TDP 1432		

Figure 1.2   Control strip from the Geneva ACC

This strip shows the identification of the aircraft (SWR665), the departure and arrival airports (LEPA and LSZH), the waypoints through which its flight plan route passes (TDP, MOL and KOR), and its (planned) flight level on entry to (FL310) and exit from (FL350) the sector.

Information contained on the control strips sometimes undergoes last-minute changes: e.g. due to a pilot request or controller instruction. This could be telephoned through from the adjacent sector, and entered manually at the receiving ACC. Once the aircraft becomes visible on the radar screen, it thus physically enters the controller's 'field of vision' and becomes a 'reality' under his control. The controller must then assimilate this arrival with the electronic data already available on the corresponding strip, temporarily 'zooming' in from the wider picture to focus on this particular aircraft, then integrate this information back into the set of aircraft under his control, assessing the associated consequences – especially for separation.

Controllers must constantly assess the consequences of each entry or exit and the evolution of the aircraft according to route, flight level and speed. For each potential risk detected, controllers mentally programme a series of solutions to avoid non-standard crossings and, at the appropriate time, issue instructions to the pilots. In response to every movement in their radar space, controllers develop extrapolation processes, forming a mental picture of what is likely to happen. Depending on their assessment of the potential risks, the controllers prioritise the order in which the conflicts are considered and dealt with.

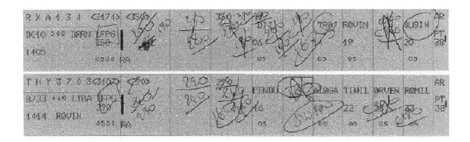

Figure 1.3   Two control strips showing complexity of annotation

On the strip, the controllers write any information that may be useful to them in monitoring the flight. All instructions given to the aircraft are noted on the strip. It is placed on a rack known as a strip rack, which the controllers organise according to vertical logic (the lowest aircraft in altitude is right at the bottom of the strip rack), followed by time logic (for two planes at the same level, the strip for the plane that is to enter the sector first will be placed lower than that of the plane that is to enter second).

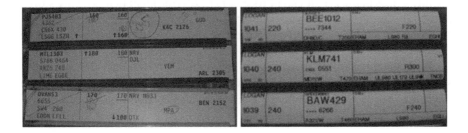

Figure 1.4   Strip racks at Geneva ACC (left) and Swanwick (right)

The strips, which are thus classified according to principles that are clear and comprehensible for the controllers, make an important contribution to building up the mental image of the developing traffic situation. At Geneva ACC, the different colours of the strip holders are also a visual aid for the controllers:

aircraft on odd-numbered flight levels are classified by yellow strip-holders, whilst those on even-numbered levels use blue strip-holders (this is not very pronounced in Figure 1.4 (left), but the upper two (blue) strip-holders can still be clearly seen to have the darker edges). London ACC also uses colour-coded strip holders, with white strips.

1.5.2   Secondary Radar and Flight Data Processing

In most of the northern hemisphere, over continental landmasses, radar is used at ACCs and airports to monitor flights. FL 600 is about the usual limit of valid performance, for both primary and secondary radar. Not many specifications are written for above FL 500, and this depends on turn rate, to a certain degree. Most commercial airports have a primary ('conventional') radar, although it is not absolutely necessary. In the UK, NATS provides a radar picture from their national service (for a fee) to many airports – mainly secondary radar data.[22] Coverage close to some airports may be poor, whereby Automatic Dependent Surveillance – Broadcast (ADS-B) becomes an option.

With secondary radar, the aircraft transponder is interrogated by a radio transmitter on the ground for a 4-digit identification 'squawk', assigned by the controller:

Controller: **'Alitalia 456, radar, squawk 2356'**
Pilot: **'Alitalia 456, squawking 2356 and requesting flight level 360'**

Taking an example flight from Hamburg to Palma, the flight does not normally change squawk. The pilot is given a code in Hamburg and keeps the same code until Palma, and *vice versa*.[23] This identity label, once passed to the controller's station, is shown alongside the aircraft's position on the radar screen.

No IFR flight can fly without first having a flight plan deposited. The radar data processor 'correlates' the transponder information with the flight plan data already stored in the Flight Data Processor (FDP) (the flight plan data are stored in each of the ATC units the aircraft is planned to fly through). Controllers may therefore speak of a 'lack of correlation' if radar (or 'status') data cannot be matched with FDP ('flight plan') data.

From the transponder, in addition to the aircraft's transponder code, the radar screen also shows (altimeter-determined) altitude and attitude (e.g. whether the aircraft is descending or climbing). The controller may thus immediately spot if

---

22   Also known as 'Secondary Surveillance Radar' (SSR). The SSR range is usually around double that of primary radar.

23   A limitation is that the transponder codes are limited in range (although a four digit-code, each digit is only in the range 0–7). This limits the availability. In order that a radar tracker can follow all the codes in his area of detection, mitigation procedures have been agreed. The so-called 'code allocation protocol' or ORCAM, is in use in Europe. In Switzerland, for example, the ORCAM area changes – which means that most flights entering and leaving the Swiss control zone from the south have to change squawk.

the altitude information from the aircraft shows any inconsistency with the cleared level. The radar antenna also senses the speed (comparing different positions after each radar return).

The controller may normally choose which data items are displayed on the screen, just as a PC user can arrange their own desktop in Windows, according to personal preference, and balancing information needs against potential clutter. Each ATC unit will define its own defaults and particular ways of operating such systems. Callsign and flight level are always displayed, and waypoints/boundaries are usually shown. In summary, labels with the following items may be used:

- squawk code;
- callsign[24] (e.g. example British Airways uses the prefix 'Speedbird');
- flight level (altitude);
- attitude;
- (planned) departure flight level;
- speed;
- flight plan data.

As a flight passes from one sector to the next, or from one controller to another, the controller instructs the pilot to switch to a new frequency. It is a basic rule of ATC that transfer of control from one controller to another must only take place if the transfer has already been agreed by direct communication between the controllers, or by a standing agreement between sectors – based on exchange of (a) letter(s) of agreement. Radar 'handover' (also known as 'handoff') is defined in CAP 493 (CAA, 2006c) as:

> Transfer of responsibility for the control of an aircraft between two controllers using radar, following identification of the aircraft by both controllers.

> The transfer of responsibility for an aircraft from one radar controller to another may be effected provided that:

> a) satisfactory two-way speech communication is available between them;
> b) the radar identity has been transferred to the accepting radar controller, or has been established by him; and
> c) the accepting radar controller is informed of any level or vectoring instructions applicable to the aircraft at the stage of transfer.

Handovers may be verbal or automated/electronic (the latter may also be referred to as a 'silent' handover). A distinction is to be made between transfer of control and transfer of communication (which are not necessarily the same thing). Transfer of communication from a controller takes place as soon as that controller has no further need for the aircraft to be on his frequency. Transfer of

---

24   Through correlation with the transponder code, from the FDP.

control starts only after the aircraft crosses the line of responsibility, as defined, for example, by the letter(s) of agreement between the adjacent ACCs.[25]

Standing agreements exist between sectors to save repeatedly agreeing the same types of transfer of control over and over again, which would become very tedious, not to mention wasteful of time and RT space. Many traffic flows are similar and predictable from day to day, and only when the handover falls outside the particular scope of the associated standing agreement, does a special case arise. The accepted minimum separation between two aircraft on the same track and at the same level are part of the negotiated procedures which will be included in the letter of agreement. A controller always tries to pass the best-sequenced aircraft to his next colleague. Around 95 per cent of transfers of control take place according to standing agreements, with the other 5 per cent normally being handled by telephone. The transfer of communication and responsibility *within* a sector is defined by local ATM procedure handbooks.

1.5.3 Controller Types and their Workstations

We now turn to look in more detail at the controller's workstation, often referred to as his 'position'. Although these vary across Europe, even from centre to centre, and from tower to tower, the general principles of operation, of workstation design, and controller interaction, are the same. Generally speaking, one frequency is manned by at least one controller. During some periods (e.g. at night, or during agreed procedural conditions – such as low traffic volumes), however, several frequencies are 'coupled' and could be worked by one controller.

Most commonly, each en-route or approach sector is manned by at least two controllers:

* the **tactical controller**[26] **(TC)**
* the **'planner'** or **planning controller**[27] **(PC)**

and, depending on traffic conditions and sector design, often with a sector (flight data) assistant and/or sector supervisor. Each of the TC–PC pair is equally responsible for the success of the prime mission: safe, orderly and efficient handling of the assigned traffic. Although each of the positions have clearly defined roles, they share a set of common responsibilities and are, in most cases, only able to handle the traffic by working as a pair. They usually sit next to each other, and thus communicate verbally.

---

25 In terms of online data interchange (OLDI), an Advanced Boundary Information (ABI) Activation message (ACT) is first sent, which is then confirmed by a Logical Acknowledge Message (LAM).

26 Also known as: (the) 'radar(ist)'/'radar executive'/'radar controller'/'executive controller'.

27 Also known as the 'radar planner' or 'coordinator'.

Figure 1.5   TC-PC pairs at Geneva ACC (left) and Swanwick (right)

The planning controller has to:

*   coordinate and approve the entry and exit of flights into the sector, and identify the aircraft that are subject to its jurisdiction (this is mostly a monitoring role, in fact). Any *ad hoc* coordinations are carried out mostly by phone, e.g. with other sectors and/or adjacent ACCs. The planner has to identify any entries into the sector where correlation is not established;[28]
*   manage the strip rack by placing the control strips on the rack as soon as they are received from the printer, plus classify the strips and keep them up to date;
*   make any flight data corrections in the FDP;
*   identify flight paths which are least likely to generate potential conflicts as flights progress in the sector;
*   monitor any additional frequencies, such as the emergency frequency.

The tactical controller is responsible for the 'front-line' RT communication with the pilots. He accepts the aircraft into the sector, monitors their progress, detects possible conflicts, issues instructions to eliminate such conflicts and achieves the exit conditions set by the planner. Generally, the tactical controller is more highly loaded than the planner, as we will explore in Chapter 3. Each controller has:

*   a radar screen;
*   a radio set for communication with aircraft[29];

---

28   How often this happens depends on the ATC unit and the level of automation. It could happen frequently, or only rarely. It may also happen if the aircraft's transponder were to be out of order, which will be announced by the previous sector, if already known.

29   To communicate with pilots, the controller can choose, depending on the prevailing rules, either to use the microphone with speakers integrated into the workstation, or to wear headphones with an integrated microphone.

- an auxiliary screen[30] (providing additional information such as weather reports, the status of systems, predicted workloads per sector and flight plan data);
- a telephone set for communication with the other positions of the same centre, or any other centre (the lines used are pre-programmed).

1.5.4   Electronic Control Strips

As an example of an increasingly automated environment, the Upper Area Centre of Geneva (above FL 245) is now working without strips. A variety of computerised lists replace the paper strip, with inputs using a keyboard and mouse. This has the two major advantages of being directly linked to the FDP (so all updates are automatic) and of being quicker to use than the manual writing out of strips. Some European ACCs are already working *entirely* with electronic strips: the writing medium of paper has completely disappeared from them.

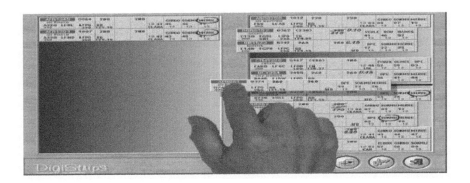

Figure 1.6   **An earlier prototype of an electronic strip rack (Guichard, 2000)**

With the increasing availability of more automated systems, and notably a less paper-based ATC system, a particular emphasis has to be placed on system redundancy and coping with degradation. For example, how will the system cope with computer failure, and how do we ensure that the controller's mental image of the situation, a vital component of operation, is not decreased in quality through increasing automation? System failure scenarios are addressed through various specific technical solutions, such as independent emergency displays.

---

30   At Swanwick (Figure 1.5), we can see that each controller has an auxiliary screen to the left of the radar display. These are locally known as 'SIS' (Support Information System) screens. Controllers can drag and drop items between the SIS screen and the radar display using the same mouse.

1.6   A Look Ahead

This chapter, the longest in the book, has laid the foundation for us to explore a number of themes. Hopefully, even the reader new to ATM, will now feel equipped to pursue our discussions as we expand upon many of these themes in the following chapters.

We have shown in this chapter how ATM has evolved from its earliest beginnings, through to the present day tasks of the controller, whose job is extremely complex and technologically driven, due to both the high demands from the users of the airspace and the great extent of automation of ATC and ATM. Nevertheless, it is perhaps surprising that the job is not yet as automated as one might expect.

We have concluded with a glimpse of the implications of an ever increasingly automated environment, a theme to which we shall return, particularly in Chapter 8, with specific regard to the human perspective, and in our conclusions. This will follow an investigation of the future prospects for air transport and the future structuring of the system, in Chapters 6 and 7, respectively.

Having now established the structure and organisation of European airspace, and looked at its operations from the controller's perspective, we will turn in the next chapter to the perspective of the airline and flight planning. In Chapter 3 we will explore the important issue of capacity, and look at the concept of 'free flight' and the delegation of separation, before examining the problems, costs and challenges of delays, in Chapter 4, and of the environmental impacts of aviation, in Chapter 5 – increasingly in the public image.

Chapter 2

The Principles of Flight Planning and ATM Messaging

Graham Tanner
University of Westminster

2.1   Introduction

This chapter discusses the principles of flight planning, the information required and the messages exchanged between the various agents. Procedures and relevant information have been drawn together from a number of sources, with a particular focus on EUROCONTROL handbooks and manuals associated with the current Central Flow Management Unit System (the suite of software known as 'CFMU 11'). An example passenger flight is used to help explain how a flight plan[1] is prepared and the processes involved in getting an aircraft airborne. Although this chapter focuses on flights operating in Europe under instrument flight rules (IFR), occasional comparisons are made with non-European flight planning and visual flight rules (VFR).

In Europe, an Air Traffic Flow Management (ATFM) service has been established to utilise capacity to the maximum extent possible while enabling a safe, orderly and expeditious flow of air traffic. The centralised service follows procedures for a regional ATFM as prescribed in ICAO's *Procedures for Air Navigation Services – Air Traffic Management* (known as 'PANS-ATM' and 'Doc 4444'), though adapted for Europe in ICAO's *Regional Supplementary Procedures* (known as 'Doc 7030'). European ATFM has evolved into the new concept of Air Traffic Flow and *Capacity* Management (ATFCM) emphasising the need to balance the management of limited capacity with the demand of increasing traffic (refer to Chapter 3 for an in-depth look at capacity issues). The terms ATFM and ATFCM are almost interchangeable in most contexts, however this chapter refers to the concept as ATFCM and to the messaging process as ATFM. There are three phases to ATFCM (set down in Doc 4444):

1) *Strategic Flow Management* occurs seven days or more before the day of operation when long-term demand and capacity matching will be planned, for example, planning for extra traffic generated by the annual winter season

---

1   Defined as 'specified information provided to air traffic services units, relative to an intended flight or portion of a flight of an aircraft' (Doc 4444).

'SKI Flow' to mainly France and Switzerland, and large events like the 2006 World Cup in Germany.

2) *Pre-Tactical Flow Management* occurs in the six days before the day of operation ('D-6' to 'D-1') during which time the strategic plan is fine-tuned in the light of updated demand as filed flight plans are received to join the earlier submitted repetitive flight plans. The ATFM Daily Plan (ADP) is published in preparation for the third phase.

3) *Tactical Flow Management* occurs on the day of operation ('D-day') with updates made to the ATFM Daily Plan as actual traffic and capacity is known. Traffic is managed through slot allocation and re-routings.

With the increased prominence placed on capacity, to get the best out of the ATM system requires cooperation between its partners – airspace users, airports, the military, and air traffic control centres, using Collaborative Decision Making (CDM) techniques (Skyway editorial article, 2005). An example of this is the 'Flexible Use of Airspace'[2] (FUA), as introduced in Chapter 1, whereby airspace is no longer designated as either 'military' or 'civil' but considered as a continuum and used flexibly on a day-to-day basis. FUA affords the following adaptable airspace structures: Conditional Routes (CDRs), available in three categories of non-permanent ATS routes, or sections thereof, which can be planned and used under specified conditions; Temporary Segregated Areas (TSAs), airspace that can be reserved for the specific use of the military; Cross Border Areas (CBAs), that are TSAs established over international boundaries; and Reduced Coordination Airspace (RCA), specifying an area of airspace implemented when Operational Air Traffic (OAT) is light or has ceased, permitting General Air Traffic (GAT) to operate outside the ATS route structure (EUROCONTROL, 2002). FUA is currently a target for enhancement, as part of DMEAN (Dynamic Management of the European Airspace Network). This is a framework partnership of aircraft operators, airports, the military, ANSPs, regulatory authorities and EUROCONTROL working to increase ATM system capacity in the short-term, until operational improvements from SESAR (refer to Chapter 7) materialise from 2010 onwards. By consolidating current initiatives in airspace design, FUA, ATFM and CDM thereby providing a coherent set of enhancements to the European operational environment, DMEAN aims to deliver an additional 10 per cent effective capacity beyond the 10–15 per cent projection outlined in existing capacity enhancement plans for 2006–10.

---

2   Originally known as the 'Flexible Use of Airspace Concept' (FUAC) when introduced in March 1996 and implemented in 30 states.

2.2   Distribution of Messages in Europe

By 1996, EUROCONTROL's Central Flow Management Unit[3] (CFMU) had taken over responsibility for the range of Air Traffic Flow Management (ATFM) services in Europe previously handled by the regional flow management units (Fournie, 2005). This development recognised that significant changes were necessary if the increasing level of delay to flights experienced throughout the 1980s was to be tackled. Before this centralisation of tactical operations, aircraft operators had to file instrument flight rules[4] (IFR) flight plans to all the Air Traffic Service Units (ATSU) concerned with the flight. Following centralisation, all such flights are filed to CFMU for processing by the Integrated Initial Flight Plan Processing System (IFPS) – the single source for distribution of IFR flight plan messaging in Europe.

At this stage it is worth clarifying the area of Europe covered by CFMU. The area included in the flight planning and message distribution service is known as the IFPS Zone (IFPZ) – this is the area for which CFMU is responsible for receiving, checking and distributing FPL messages and includes almost 40 states (slightly larger than the membership of EUROCONTROL). For states at the geographical periphery of the IFPZ, it may be case that not all their Flight Information Regions (FIRs) are included, for example Shanwick airspace (EGGX) is not within IFPZ even though controlled by the Scottish Oceanic and Shannon Area Control Centres (ACCs; UK and Ireland[5]). Airspace outside IFPZ that receives copies of flight plans from CFMU is known as the Flight Plan Messages Copy Distribution Area ('FPM Copy'). The airspace covered by the combination of IFPZ and FPM Copy constitutes the 'CFMU Area'. Incidentally, the CFMU Area includes other operational function areas such as the Air Traffic Flow Management Area (airspace subject to ATFM measures) and airspace subject to en-route charges calculated by EUROCONTROL's Central Route Charges Office (CRCO), but these are within the areas already described. The number of states included in the CFMU Area has increased over the last decade, Figure 2.1 illustrates the situation in 2007 (plotted using SkyView2) (EUROCONTROL, 2007c).

Beyond the CFMU Area boundary, the airspace of cooperating states may also be subject to ATFM measures, known as the ATFM Adjacent Area (ATFM ADJ).

2.2.1   Integrated Initial Flight Plan Processing System (IFPS)

The Integrated Initial Flight Plan Processing System (IFPS) receives, processes and delivers IFR flight plan data within the IFPS Zone. IFPS is split between

---

3   Established following a meeting held in Frankfurt (1988) between European Civil Aviation Conference (ECAC) Ministers of Transport.

4   Flights under the rules governing instrument meteorological conditions.

5   Shanwick traffic is managed by the Oceanic Area Control Centre (ACC) located in Prestwick (Scotland); radio communication is made via Shannon Aeradio, also known as 'Shanwick Radio', located near Shannon (Ireland).

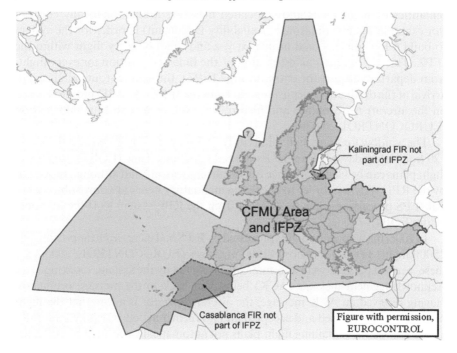

Kaliningrad FIR not
part of IFPZ

CFMU Area
and IFPZ

Casablanca FIR not
part of IFPZ

Figure with permission,
EUROCONTROL

Figure 2.1   CFMU Area and IFPS Zone (IFPZ) in 2007

two functionally identical units, the FP1/RPL Section[6] in Haren (Brussels) and the FP2 Section[7] in Brétigny-sur-Orge (just outside Paris). Flight plans are sent to both sites and the workload is then shared between them. FP1/RPL processes messages for flights departing airports located in northern Europe as well as repetitive flight plans (RPLs, handled via the RPL Unit first) and air filed flight plans (AFIL, see later section). FP2 processes messages for flights departing from airports around the rest of Europe and countries surrounding the IFPS Zone (intending to fly into Europe). In addition to this separation of processing, should one unit fail, the remaining operational unit can take over responsibility for all processing (EUROCONTROL, 2006f).

As previously noted, IFPS is the sole source for distributing IFR flight plan messages within the IFPZ. (In the case of mixed IFR/VFR flights and mixed GAT/OAT[8] flights – only the IFR and GAT parts of the flight plan should be filed). This is to ensure that aircraft operators, airports, air traffic control centres and CFMU all work with the same data. Once received by IFPS, the syntax and

---

6   FP1/RPL Section was previously known as IFPU1.
7   FP2 Section was previously known as IFPU2.
8   As introduced in Chapter 1, General Air Traffic (GAT) covers civilian flights 'conducted in accordance with the rules and procedures of ICAO and/or the national civil aviation regulations and legislation' (CAA CAP 694), whereas Operational Air Traffic (OAT) covers military flights.

semantics of flight plans and associated messages are automatically checked (for example, against the Route Availability Document) in order for the system to build a four dimensional profile (x, y, z and time) of every flight within the IFPZ. This 4D profile is an estimation of the time and position for every flight, from departure airport (or entry into the IFPZ), through en-route airspace, to arrival at the destination airport (or exit from the IFPZ). This 4D profile is based on the aircraft performance and filed route, level, speed and time information (EUROCONTROL, 2007b).

The window for filing flight plans to IFPS for processing is between three and 120 hours (five days) before the Estimated Off-Block Time (EOBT). Repetitive flight plans can be filed in advance of the current season,[9] and although processed by the RPL system in order to check the mandatory items, are only subjected to full IFPS processing when transferred from the RPL system to IFPS, 20 hours before EOBT.

In 2006, the number of flights processed by IFPS had increased to approximately 25000 flights per day – almost 40000 messages (EUROCONTROL, 2006i). Of these, over 80 per cent were automatically accepted by the system, resulting in an Acknowledgement message (ACK) being sent back to the message originator. Having received an ACK message, the aircraft operator is aware that the flight plan has been processed and accepted and no further action is required until slot allocation. Of the remaining flight plans not automatically accepted by IFPS, the vast majority are accepted following manual intervention by IFPS staff – such intervention identified by a Manual message (MAN) returned to the message originator, indicating that no action is required. If manual intervention is successful an ACK message is sent back, however, if the message remains invalid a Rejection message (REJ) is returned – compelling the message originator to act before re-submitting the flight plan (EUROCONTROL, 2007b). 'Until an ACK message has been received by the message originator, the requirement to submit a valid FPL for an IFR/GAT flight intending to operate within the IFPS Zone will not have been satisfied' (UK AIP, ENR 1-10-11, paragraph 3.6.2.7 (CAA, 2007b)). IFPUV (IFPS validation system) offers a facility to automatically test the validity of FPL syntax before submission to IFPS – giving a reply informing whether the content of a FPL is valid, or if not, why it was rejected. IFPUV is not connected to the operational IFPS, so messages received are neither distributed nor stored.

Each flight plan in the IFPS database is assigned a unique identifier ('IFPLID') which is then used with messages associated with each flight. IFPS distributes accepted flight plan messages to all the Air Traffic Service Units (ATSUs) associated with the movement, including those outside IFPZ that have been included using the re-addressing function (see later), and to the Enhanced Tactical Flow Management System (ETFMS) in readiness for any flow regulations. Once processed by IFPS, the ICAO-formatted filed flight plan messages are distributed in ADEXP[10] format (repetitive flight plans have to be submitted in this format,

---

9   There are two seasons: 'summer' from the end of March and 'winter' from the end of October.
10   ATS Data Exchange Presentation.

whereas filed flight plans can be submitted in either ICAO or ADEXP format). ADEXP format is discussed with examples in a later section.

2.2.2   *Submission of Flight Plans to IFPS*

There are a variety of tools available for generating and submitting IFR flight plans to IFPS, however, there are only two networks used for transmitting the messages – AFTN (Aeronautical Fixed Telecommunications Network) and SITA (originally Société Internationale de Télécommunications Aéronautiques). The CFMU Network and the Internet are also used for message transmission between CFMU and aircraft operators (or their agents) by way of the various CFMU applications, but flight plans are not filed using them. 'Direct filing' describes the procedure when aircraft operators file their own flight plans to IFPS (and any other non-IFPS states) rather than forwarding them to the Air Traffic Service Unit (ATSU) at the departure aerodrome, in order for them to file on the aircraft operator's (AO) behalf. All this applies to filed flight plans (FPLs). Repetitive flight plans (RPLs), being data files rather than messages, should be submitted to the RPL Section either by e-mail or by SITA telex/SITATEX (SITATEX is a message application).

AFTN is a ground-to-ground civil aviation communication network dating back to the 1950s and conforms to standards laid out by Annex 10 to the ICAO *Convention on International Civil Aviation*[11] (Annex 10 Aeronautical Telecommunications – specifically Volume III, Communication Systems). All significant ground points around the world, such as airports, air traffic control centres and meteorological offices, are linked by this network enabling messages to be exchanged through all phases of flight. Almost every country in the world has a node on the network, resulting in several hundred nodes exchanging messages on links of different types and speeds using 'store-and-forward' procedures. Europe is a busy part of the AFTN, handling not only the large volume of message traffic within Europe but also considerable transit traffic – a major node might handle 100 000 AFTN messages per day. Most European nodes have now been upgraded with modern technology to create the CIDIN (Common ICAO Data Interchange Network) in order to increase capacity. The CIDIN supports AFTN, and the network is still referred to as AFTN (EUROCONTROL, 2007e).

The SITA network was founded in 1949 originally as a cooperative that brought together existing airline communications facilities to create a shared infrastructure. Services were initially available to IATA members only, but the network has expanded to now serve over 500 airlines, Computer Reservation Systems (CRSs), airports, other aviation companies and governments around the world (SITA, 2006).

The AFTN and SITA networks are closed networks – in order to send and receive messages over them, users need to be registered (assigned with a unique address). AFTN addresses consist of eight characters and are hierarchically

---

11   The 1944 *Convention on International Civil Aviation*, commonly referred to as the 'Chicago Convention'.

structured with the first four characters consisting of the ICAO station code (the location indicator[12]); characters five to seven being the ICAO designator of aeronautical authority or company; the final character signifying the department or division, else the letter 'X' is used (ICAO, 2001). SITA addresses consist of seven characters, the first three characters being the IATA station code. Refer to Table 2.1 for examples of AFTN and SITA addresses using the two IFPS units. As we will see in the next section, an address is required when submitting a flight plan in order to disclose the message originator (that is, the aircraft operator), as well as the address of the intended recipient(s).

Table 2.1   Addresses of the IFPS units

IFPS units		AFTN	SITA
	FP1	EBBDZMFP	BRUEP7X
Haren	RPL	–	BRUER7X (via SITATEX)
	IFPUV	*EBBDZMFV*	*BRUEY7X*
Brétigny-sur-Orge	FP2	LFPYZMFP	PAREP7X

As mentioned, in addition to the AFTN and SITA networks, the Internet and CFMU Network are also available for message transmission through different CFMU applications. The CFMU Network is a secure high speed TCP/IP Extranet supplied by SITA (though separate from the previously described SITA infrastructure) through which the CFMU Human Machine Interface (CHMI) connects users interactively with real-time information. Around 1700 aircraft operators, 100 Flow Management Positions (FMPs) and 250 ATS Units access CFMU systems this way (EUROCONTROL, 2006f). There are several different CHMI services available to suit the requirements of these users: CIAO for Aircraft Operators, CIFLO for Flow Management Positions and CIREN for ENV (Environment) Coordinators. Users accessing CFMU systems via the Internet can make use of the CFMU Internet Application (CIA) and the CFMU Interactive Reporting (CIR) applications (EUROCONTROL, 2007d). The CIA and CHMI applications replace the older 'CFMU Terminal' (RTA/RCA – Remote Terminal Access/Remote Client Application), and enable users to access the Enhanced Tactical Flow Management System (ETFMS) as well as send and receive ATFM messages. The same functions are available to both the CIA and CHMI applications, though the latter has a graphical interface for displaying maps. The CIR application provides access to archived data and is used to generate customised statistics (not messaging).

So far this section has mainly discussed messaging networks. As mentioned, there are a variety of software applications available to flight planners, serving the requirements of single aircraft operators through to the largest carriers, planning

---

12   The four character ICAO location indicator is itself a hierarchical code.

thousands of movements on a daily basis. There are many online tools that will generate flight plans, though generally these are aimed at VFR flights and are not suitable for submission to IFPS. Some of the widely used flight planning suites such as 'Lido Operations Center' ('Lido OC', Lido originally standing for 'Lufthansa Integrated Dispatch Operation'), 'DispatchPro' and 'RODOS Flight Planning' ('ROute DOcumentation System') are able to automatically generate and distribute flight plans for all such operations of an airline. Depending on the level of functionality paid for, such systems use up-to-date weather and aeronautical information (for example, current routings and the various daily messages such as NOTAMs and AIMs – see later) combined with 'company' information (such as the differing performance characteristics of individual aircraft in the fleet) to calculate optimal flight plans for the day's operations. These systems will generate internal operational flight plans containing greater detail than is filed as an ICAO flight plan, enabling dispatchers to decide between a number of scenarios (for example, predicted fuel burn compared with flight time, routes and the associated en-route charges). Once flight plans have been filed via the AFTN or SITA networks, some of these systems are able to receive and automatically act upon Slot Allocation Messages (SAMs), and if the aircraft are suitably equipped, communicate with the flight crew via datalink.[13]

Whilst discussing flight planning software, the production of repetitive flight plans (RPLs) should not be overlooked. CFMU provide aircraft operators with an application for the preparation of RPLs – the CFMU RPL Input Application (PRFPL). This ensures that RPL list files are submitted to the RPL Unit electronically (e-mail or SITA telex/SITATEX) in the correct IFPS RPL format. Other applications can be used (flight planning suites can manage RPLs), but the submitted list must be in IFPS RPL format. Aircraft operators should no longer submit hard copy RPLs by post or fax, which in any case carries the risk of errors being introduced as the lists are transcribed (EUROCONTROL, 2001c).

The proportion of flight plans generated from RPLs was approximately 45 per cent of all flight plans submitted to IFPS in 2004. The contribution that RPLs make is decreasing each year, even taking account of the growth in the airline sectors most likely to use them (regional and eastern European airlines), as more aircraft operators make use of the benefits of filing flight plans on a daily basis (EUROCONTROL, 2005a). There are of course positives for AOs in utilising RPLs. For example, there may be a reduction in workload by submitting a season's worth of flight plans at once, or if a failure occurred to the AFTN or SITA networks, the day's flight plans are already in IFPS (EUROCONTROL, 2007b).

Certainly aided by the levels of functionality offered by modern flight planning software, by FPL rather than RPL, aircraft operators can select routings based on knowledge acquired at a pre-tactical or tactical level. In addition to taking into account the meteorological situation, routes can be filed that include available

---

13   Datalink is an automatic and manual messaging system which provides crews in suitably equipped aircraft, on the ground or airborne, with real-time ground communications.

Conditional Routes, specifically CDRs 2 released by the daily Conditional Route Availability Messages (CRAMs).

2.3   Planning a Flight and the Information Required

Although there are situations when a flight plan does not need to be filed (such as 'booked out'[14] VFR flights), this section examines the typical content of flight plans filed to CFMU. That is, all IFR/GAT flights intending to operate partially, or wholly, within the IFPS Zone (refer to Figure 2.1) and subject to ATFM measures. Incidentally, a written flight plan may be volunteered for any flight even if such filing is not required, may still be filed by the pilot or operator to the Air Traffic Service Unit (ATSU) at the departure aerodrome.

2.3.1   Aeronautical Information

IFR and VFR flight planning depends upon the provision of Aeronautical Information Services (AIS). AIS is defined by EUROCONTROL's European Air Traffic Management (EATM) programme as 'a service provided for the collection and dissemination of information needed to ensure the safety, regularity and efficiency of air navigation. Such information includes the availability of air navigation facilities and services and the procedures associated with them, and must be provided to flight operations personnel and services responsible for flight information service' (EUROCONTROL, 2005b). The provision of AIS is laid out by Annex 15 of the Chicago Convention. Two other annexes also have relevance to this discussion on flight planning – Annex 3 outlines meteorological services and Annex 4 aeronautical charting. Meteorological information is not only important to aircraft in flight but also in terms of planning the route, for example to make use of favourable winds and conserve fuel. Aeronautical charts are standardised (if 'ICAO' is in the title) to form a series that serve different purposes, including flight planning, and some of which make up publications under the AIS umbrella.

Focusing on AIS, Annex 15 specifies the aeronautical information to be made available by states, the primary source of which is the Aeronautical Information Publication (AIP). The AIP is a comprehensive reference document of procedures, descriptions, charts and other information presented by states in a standardised format in three parts – general ('GEN'), en-route ('ENR') and aerodrome ('AD'). Some information within the AIP is valid for years (such as the facilities available at an airport), whereas other information may require updating more often, or on a cyclical basis. In order to synchronise changes a system of predetermined 'effective dates' every 28 days is in place (always on a Thursday), known as the AIRAC (Aeronautical Information, Regulation and

---

14   The ATSU at the departure aerodrome is informed of the intention to make a flight; however the flight details are not transmitted to any other ATSU.

Control) cycle[15]. In addition to updates made to the AIP each AIRAC, the AIP can be updated by Amendments ('AMDT'); by Supplements ('SUPs') for long term temporary changes; by Notices to Airmen ('NOTAMs') and Pre-flight Information Bulletins ('PIBs') for short-term announcements such as runway closures; and by Aeronautical Information Circulars ('AICs') for administrative information. Electronic versions of the documents that make up the AIP are now widely available and accompany the paper-based editions. States may offer additional sources of aeronautical information, for example the *UK and Ireland Standard Route Document* ('SRD'), published each AIRAC, lists preferred routings within these countries' airspace.

CFMU collate information from AIPs, NOTAMs and other documents (such as military) into the European AIS Database ('EAD'), enabling aircraft operators to access a single source of aeronautical information for the ECAC area. EAD went live in June 2003 and is the world's largest centralised AIS reference database, offering different levels of functionality to suit user requirements, such as EAD Basic and EAD Pro (EUROCONTROL, 2006j). Another key CFMU publication is the Route Availability Document ('RAD'), the 'sole-source-planning document which integrates both structural and Air Traffic Flow and Capacity Management (ATFCM) requirements geographically and vertically' (EUROCONTROL, 2006k). The RAD consists of two parts that list the restrictions on routes applicable to IFR. Part I contains separate annexes with restrictions for each state, the North Atlantic and military traffic, along with four appendices that define the RAD (in general terms), area definitions (for example FIRs, airport groups), city-pair level capping (between the FIRs and airport groups) and direct route limitations (in each state). Part II lists all the restrictions across Europe, and should correspond with Part I. Restrictions can be: 'inclusive' in that traffic must meet all the conditions for the restriction to apply (for example, departing airport group*1*, with destination airport group*2* and above a particular flight level); 'exclusive' in that traffic need only meet one condition for the restriction to apply; or 'compulsory' in that traffic must fly this route. Examples of using the RAD to build a route appear later in this chapter. The RAD is updated each AIRAC and is available as a series of stand-alone documents, as well as through EAD.

Other aeronautical information message types published by CFMU include the Conditional Route Availability Message ('CRAM'), listing which CDRs are open and which are closed the day before operation; the ATFM Notification Message ('ANM'), listing planned ATFM measures the day before operation; and the ATFM Information Message ('AIM'), listing instructions and information about current ATFM measures on the day of operation. These latter two message types are ATFM messages (see later) though are pertinent at the flight planning stage.

The AIS provision discussed so far has introduced the direct sources of aeronautical information. However such information is also available to aircraft operators through suppliers such as Jeppesen – reformatted to suit company requirements if necessary. As discussed earlier, flight planning software is able

---

15   The AIRAC cycle was adopted by ICAO in 1964.

to process and use electronic aeronautical information in order to automatically generate flight plans, avoiding hand copies.

There are links to many of these sources in the 'European ATM: a Compilation of Web-based Information Resources' at the back of this book.

2.3.2   Filed Flight Plan Explained

To assist with the explanation of the information contained in a filed flight plan, an example is referred to (Figure 2.2). This is the flight plan for a fictional scheduled weekday service in May between London Heathrow and Frankfurt.

```
(FPL-GST123-IS
-B738/M-SDRPWY/S
-EGLL1500
-N0444F330 DVR6J DVR UL9 KONAN UL607 SPI UT180 DITEL T180
 POBIX/N0414F230 T180 OSMAX OSMAX2E
-EDDF0057 EDFH
-EET/EGTT0008 EBUR0017 EDVV0041 EDUU0041 EDGG0047 REG/GGGJL
 SEL/MITJ OPR/TANNER AIRWAYS DOF/070504 RMK/TCAS)
```

Figure 2.2   Example flight plan

The example flight plan is presented in 'ICAO format', with the message fields structured following rules set out by Doc 4444. These rules apply to all standard ATS messages around the world, such as filed flight plans (FPLs), departure (DEP) and arrival (ARR) messages,[16] though not to the message types unique to IFPS (such as the Operational Reply Messages: ACK, MAN and REJ). These standard ATS messages are composed with a regular order of data selected from the 22 (sequentially numbered) available message fields, also known as 'items'. Some of these data fields are currently not used – such as items 1 and 2 – hence item 3 which describes the message type (for example 'FPL') is the first message field. In the case of filed flight plan messages, items 3, 7, 8, 9, 10, 13, 15, 16 and 18 are required (described shortly). For comparison, arrival messages contain items 3, 7, 13, 16 and 17.

Figure 2.2 illustrates how flight plan data is filed electronically. These are the same data that could be written (or printed) on an ICAO model flight plan form (Appendix 2, Doc 4444) if manual filing was necessary. Reproductions of the ICAO model flight plan form are usually published by Air Navigation Service Providers (ANSPs), for example in the UK it is known as Form CA48/RAF F2919. In the US it is known as the 'International Flight Plan' or FAA Form 7233-4, distinguishing it from the (different) 'domestic' FAA Form 7233-1

---

16   Departure and arrival messages are transmitted by the ATSU following the take-off and landing of aircraft for which flight plan data have been distributed.

format (note that flights over land between the US and Canada/Mexico may be filed as if domestic).

It is important that a filed flight plan conforms to the ICAO format[17] as even small errors such as an extra space or an expected character omitted could result in a FPL being delayed whilst manual intervention takes place, or possibly rejection by IFPS. The beginning and end of a message contents are indicated by '(' and ')' respectively – the 'Start-of-ATS-Data Signal' and 'End-of-ATS-Data Signal' (ICAO, 2001). Information outside these brackets constitutes the message envelope and is required for transmission. For example, messages sent via the AFTN (discussed earlier) begin with the code 'ZCZC' and end with 'NNNN' after the final ') – remnants of the era when messages were transmitted using holes punched in tape.[18] The start of the message envelope contains information concerning message priority, filing time, the address to where it is being sent (as well as copies to additional addresses) and the originator address. As flight plans for IFR/GAT flights are filed to both IFPS units, the AFTN address of FP1 and FP2 are required (when messages are sent via AFTN), hence the address line would be 'FF EBBDZMFP LFPYZMFP'. The next line in the message envelope contains the filing time and originator address (discussed earlier). The filing time is disclosed using six digits representing the submission date and time, for example '030900' signifies a flight plan filed at 0900 on the third day of this month (note the use of leading zeros). There can be no ambiguity with the date of filing since FPLs are submitted to IFPS no more than five days before EOBT. If required, the last line[19] before the start of the flight plan message, prefixed with 'AD', contains a list of addresses to where copies of the FPL should be sent by IFPS for flights exiting (or entering, if filed from outside) the IFPS Zone. Note that this re-addressing function is only available for AFTN addresses (EUROCONTROL, 2007b).

Before discussing the actual flight plan data between the brackets it is necessary to highlight the general conventions used when completing the required fields (more specific conventions such as indicating the planned altitude or flight level are discussed for the relevant item). Block capitals are used throughout and all clock times are UTC (Coordinated Universal Time or 'Zulu', that is, Greenwich Mean Time), with elapsed times in four digits as hours and minutes (leading zeros are used when appropriate). ICAO location indicators are always used to specify airports. Each item is delimited by a '-' hyphen (the 'Start-of-Field Signal'). Elements within a field that require separation are done so using an '/' oblique stroke or a space (ICAO, 2001).

---

17   IFPS accepts flight plans submitted in ADEXP format as well as ICAO format; ADEXP format is discussed in more detail later.

18   The distinctive pattern of holes produced by 'ZCZC' and 'NNNN' made identifying the start and end of messages a straightforward task.

19   Or last lines if the more than seven AFTN addresses are required, as this is the maximum that can appear on a single line.

Filed flight plan: items 3 (message type), 7 (aircraft identification) and 8 (flight rules and type of flight)

```
(FPL-GST123-IS
```

Item 3 is the first message field and specifies the type of message being transmitted – a filed flight plan (FPL) in the context of this chapter. This item consists of a three character code and is present in all standard ICAO ATS messages. In addition to 'FPL', other examples of message type include: 'CHG' (modification); 'DLA' (delay); 'CNL' (cancellation); 'DEP' (departure); 'ARR' (arrival) and 'EST' (estimate) messages.

Hyphens are used to separate item 7 from items 3 and 8 as these fields share the same line. Item 7 consists of the aircraft identification indicating the radio call sign (see Chapter 1), either by way of the aircraft registration markings or as in this case, 'GST123', the ICAO telephony designator[20] of the aircraft operator followed by the flight number (EUROCONTROL, 2007b). 'GST' happens to be the designator for flights operated by the fictional Tanner Airways, though if this flight was operated by British Airways it would be identified as 'BAW123'. The aircraft identification can be between two and seven characters, and is the first of four fields within an FPL classified as a 'key field' by IFPS. Key fields contain flight plan information that cannot be changed once filed – any attempts to do so will be rejected by IFPS. As well as aircraft identification, the other key fields are departure aerodrome, destination aerodrome and date of flight. To change any of these four key fields requires the filed flight plan to be first cancelled before re-filing a new one.

IFPS can process additional aircraft identification that includes the Secondary Surveillance Radar code (SSR) for the flight. If included, '/A' and the four digit SSR code follow the aircraft identification (for example, 'GST123/A2325' for this example flight), up to a maximum of 13 characters.

Item 8 contains the flight rules and type of flight (one character each), in this case 'IS'. As previously discussed, IFPS only processes the plans for flights to be conducted under instrument flight rules (IFR) within the IFPS Zone – flights filed under visual flight rules (VFR) will be rejected. There are four possible codes to describe the flight rules: 'I' for IFR only; 'V' for VFR only (not applicable with IFPS); 'Y' for flights starting off under IFR before changing to VFR; and 'Z' for flights starting off under VFR before changing to IFR. When a change in rules flown is to occur ('Y' or 'Z'), the point in the route where the change is planned must be specified in item 15 (route).

The second part of item 8 indicates the type of flight, from five possible codes: 'S' for scheduled; 'N' for non-scheduled; 'G' for general aviation (GAT); 'M' for military; and 'X' for all other flights. Incidentally item 8 is not required for repetitive flight plans as these are always processed as scheduled IFR flights (unless specified to the contrary in a remark in item 18).

---

20   ICAO's 'Designators for Aircraft Operating Agencies, Aeronautical Authorities and Services' (Doc 8585) is available online at www.eurocontrol.int/icaoref.

Filed flight plan: items 9 (number, type of aircraft and wake turbulence category) and 10 (equipment)

```
-B738/M-SDRPWY/S
```

In the sample flight plan the first part of item 9 is legitimately missing, as the number of aircraft covered by the flight plan is only required if there are more than one aircraft. If, for example, there were two aircraft intending to fly, then '02B738' would be filed (the leading zero is required as two digits are expected). Note that formation flights are not appropriate in many situations – civil formation flights are not permitted in RVSM airspace.

The type of aircraft is specified using the assigned ICAO aircraft type designator,[21] and can be between two and four characters. As the example flight plan is for a flight operated with a Boeing 737-800, the ICAO designator 'B738' is used. In instances where a designator has not yet been assigned for the planned aircraft, or more likely, when aircraft of different types are to be flown in formation (the number of aircraft must be specified) the code 'ZZZZ' is used instead. In such cases, additional information is required in item 18 using the sub-field TYP (type) to specify the generic aircraft type,[22] or number of each type of aircraft in the formation.

Following the ICAO aircraft type designator and the oblique stroke, is the wake turbulence category. Wake turbulence occurs in flight from the wake vortices generated from an aircraft's wing tips, with the strength of the turbulence related to the mass of the aircraft. This results in the need for ATC to keep a minimum separation between aircraft based on their maximum certificated take-off mass (known as the maximum take-off weight, MTOW). The MTOW is categorised so that an aircraft will fall into one of three available groups: heavy ('H') includes all aircraft with a MTOW of 136000 kg or more; medium ('M') for aircraft over 7000 kg but less than 136000 kg; and light ('L') for aircraft of 7000 kg or less. (Note that a new heavy ('J') *may* be introduced for the Airbus A380.) The Boeing 737-800 has a medium wake turbulence, with a MTOW of 79010 kg (Boeing, 2007.) IFPS uses performance data for the filed aircraft type to calculate the flight profile, and checks that the wake turbulence category given is appropriate.

Item 10 (after the hyphen) lists the equipment carried – 'SDRPWY/S'. This covers the serviceable radio communication (COM), navigation (NAV) and approach aid equipment, and after the '/', the serviceable surveillance equipment. Each character identifies equipment or a range of equipment carried.

With COM/NAV/approach aid equipment, if no such equipment is present, or the condition is unserviceable, then 'N' is used. If standard equipment is carried – considered to be VHF RTF (radiotelephony), ADF (Automatic Direction Finder), VOR (VHF Omnidirectional Range) and ILS (Instrument Landing System) – then

---

21   ICAO's 'Aircraft Type Designators' (Doc 8643) is available online at www.eurocontrol.int/icaoref and www.icao.int/anb/ais/8643/.

22   Generic aircraft types are: turbo-jet (TJJJ); turbo-prop (TPPP); single engine (SEEE); and multi-engine (MEEE).

'S' is used instead. Additional equipment that is serviceable is also listed. Note that certain types of equipment require extra information specifying in item 18, for example, if datalink ('J') is serviceable then the DAT (datalink capability) sub-field is used to describe the type available (such as satellite datalink). As most letters in the alphabet have been assigned to describe COM/NAV/approach aid equipment, attention will be focused on the equipment listed in the example flight plan 'SDRPWY'.

Since 'S' covers the standard equipment – there is no need to additionally list 'V' (VHF RTF), 'F' (ADF), 'O' (VOR) or 'L' (ILS). The availability of certain equipment is a requirement in some states and is specified in their AIPs, or even Europe-wide for certain types of airspace. The presence in item 10 of 'D', Distance Measurement Equipment (DME), is a requirement for flights for example in the UK's controlled airspace alongside the standard equipment (UK AIP, GEN 1-5-3, paragraphs 1.2.2-1.2.3 (CAA, 2007c)). 'R' refers to the availability of RNP (Required Navigation Performance) type certification, and 'P' the availability of P-RNAV (precision area navigation system) – refer to Chapter 1 for a description of area navigation (RNAV).

The presence of 'W', indicating the aircraft carries Reduced Vertical Separation Minima (RVSM) equipment, and 'Y', 8.33 kHz channel-spacing capable radio are Europe-wide requirements. RVSM was introduced to 41 states in 2002, and reduced the vertical separation to 1000 ft between FL 290 and FL 410, thereby creating six new flight levels. All flights planning to operate in the RVSM area must be fitted with suitable equipment. However, 'W' must be present in the item 10 equipment list if the aircraft is carrying such equipment even if the flight is not planned to operate between FL 290 and FL 410. Non-RVSM approved aircraft and VFR movements must file above or below the RVSM area, though exceptions are made for 'state aircraft'.[23] From March 2007, it has been mandatory for aircraft operating above FL 195 to be equipped with 8.33 kHz channel-spacing capable radio, though states may grant exemptions within their areas of responsibility (in addition to exemptions for state aircraft). Indeed, a number of states within the IFPS Zone are yet to implement 8.33 kHz channel spacing operation (EUROCONTROL, 2007b).

Moving on to the second part of item 10, serviceable surveillance equipment must be described by either one or two characters. The first character identifies the mode of SSR available, or 'N' if no such equipment is present. The second character identifies the automatic dependent surveillance (ADS) equipment carried – an on-board system that provides aircraft identification and 4D data (among others) via datalink – if present the code 'D' is used. In the example FPL, only one character ('S') is listed, identifying that 'Transponder-Mode S, including both pressure-altitude and aircraft identification transmission' is carried (ICAO, 2001). Other codes include 'A', 'C', 'X', 'P' and 'I' to describe the various SSR transponder modes, or 'N' to identify the absence of such equipment.

---

23   State aircraft are used in military, customs and police services.

Filed flight plan: item 13 (departure aerodrome and Estimated Off-Block Time)

```
-EGLL1500
```

Item 13 contains another of the unchangeable key flight plan fields in the aerodrome of departure, in addition to the Estimated Off-Block Time[24] (EOBT) at this aerodrome, forming a string of eight characters. The filed departure aerodrome consists of the four character ICAO location indicator[25], 'EGLL' being the code for London Heathrow. As mentioned earlier, the ICAO location (also known as the 'station') indicator is a hierarchical code with the first character relating to the part of the world ('E' for northern Europe), the second identifying the country ('G' for the UK) and the final two characters specifying the location ('LL' for London Heathrow). If no location indicator exists, then the code 'ZZZZ' is used, with the name of departure aerodrome included in the sub-field DEP, in item 18. There is a third possible code that can be used here – 'AFIL' – to distinguish air filed flight plans submitted to IFPS by ATS Units on behalf of aircraft already airborne. As with the use of 'ZZZZ', air filed flight plans require supplementary DEP information in item 18, however, instead of specifying the EOBT, the estimated or actual time over (ETO or ATO) is provided for the estimate point given in the filed route (EUROCONTROL, 2007b).

The Estimated Off-Block Time is comprised of four digits expressing the time in Coordinated Universal Time. This is not the published scheduled departure time, but rather the true time the aircraft operator expects the aircraft to be ready to depart (see also Figure 4.1, in Chapter 4). In this example the filed EOBT is '1500', that is, 1600 local time (British Summer Time). The scheduled departure time which passengers work to may be different. IFPS calculates the flight profile from the given EOBT (see later).

Filed flight plan: item 15 (route)

```
-N0444F330 DVR6J DVR UL9 KONAN UL607 SPI UT180 DITEL T180
POBIX/N0414F230 T180 OSMAX OSMAX2E
```

Item 15 consists of three parts that specify the initial cruising speed, initial cruising level and a description of the route. The initial cruising speed is the true air speed[26] (TAS) for the first, or entire, cruising portion of the flight. The speed can be expressed in either knots as in the example, by the use of 'N' followed by four digits, or as kilometres per hour with a 'K' and four digits, or as a Mach number to the nearest hundredth by way of 'M' and three digits (that is, M082

---

24   Defined as 'the estimated time at which the aircraft will commence movement associated with departure' (Doc 4444).

25   ICAO's 'Location Indicators' (Doc 7910) is available online at www.eurocontrol. int/icaoref.

26   The true air speed is the actual speed of an aircraft after correction for temperature and air density.

expresses Mach 0.82). Planned changes of speed are stated at the point in the filed route.

The initial cruising level for the first, or entire, cruising portion of the flight follows the speed without any spaces. As with speed, there are different ways of expressing the cruising level: the flight level ('F' and three digits) as in the example; the altitude in hundreds of feet ('A' and three digits); the standard metric level in tens of metres ('S' and four digits); the altitude in tens of metres ('M' and four digits); or for VFR not intending to fly at a particular cruising level, 'VFR' (though inappropriate in the context of filing to IFPS). The cruising level can be a maximum of five characters, with leading zeros used when appropriate. Planned changes of level must be stated in the route.

Taking the start of the example flight plan, 'N0444F330', once the top of climb has been reached the aircraft will be cruising at FL 330 whilst travelling with a true air speed of 444 knots. IFPS will accept cruising speed and level information presented in any of the units described above, enabling aircraft operators to file with their preferred measurement unit. Incidentally, IFPS automatically converts from one unit to another when appropriate. For example, any metric levels filed in a route are converted to flight levels for processing as these are required for profile calculation, though in this situation no conversions are made to the actual flight plan. Note that some countries around the world require such information in a prescribed format, for example, kph and metric cruise levels are used in the Russian Federation. The third part of item 15 describes the intended route and changes to the level, speed or flight rules that are planned to be made en-route.

The rest of this section outlines the route description in general terms, and then focuses in more detail on the route selected in the example flight plan (plotted in Figure 2.3), including the sources of information used to construct the route. For further information about the terms used, such as SIDs, STARs and waypoints, please refer to Chapter 1.

The flight plan route consists of a sequence of points and airways between the departure and destination aerodromes (or from the points of entry to exit of the IFPS Zone). Note that this section is concerned with flights planning to file along designated ATS routes (airways). Planning for flights outside these designated routes involves either specifying the points not more than 30 minutes' flying time or 200 NM apart, or defining the tracks along which flights intend to operate in a predominantly east-west (for example, North Atlantic flights) or north-south direction.

The simplest possible route with the least number of constituent parts between the two aerodromes is a direct (DCT) route, if available, and if within the maximum[27] DCT length – for example the maximum DCT in the UK's London and Scottish UIRs is 30 NM (EUROCONTROL, 2007h). Direct routes are used when the departure aerodrome is not on, or connected to, the airway, or when a Standard Instrument Departure (SID) is not available. A DCT can also be used to reach the destination aerodrome from an airway if, for example,

---

27   The maximum direct route length within a specified airspace is declared by each state.

a Standard Terminal Arrival Route (STAR) is not available. Most flight plans filed to IFPS, however, are able to include a SID and a STAR in the route (some countries may require DCTs rather than SIDs and STARs). SIDs and STARs however, are not used with repetitive flight plans. With RPLs, the connecting point between the SID and airway, and the connecting point between the airway and STAR are used – IFPS automatically inserting the most suitable SID and STAR when processing.

Waypoints may indicate positions at which aircraft join, change and leave airways, and as previously discussed in Chapter 1, are referred to in different ways. Waypoints and navaids are coded designators consisting of two to five characters. In cases where no designator has been assigned to a point, then either the geographical coordinates or the bearing and distance from a navigational aid are used instead. If geographical coordinates are used, they can be expressed as degrees only, for example '46N078W' (seven characters, with leading zeros) or as degrees and minutes, for example '4620N07805W' (11 characters, with leading zeros). If a bearing with a distance from a navigational aid is used (frequently used in the US), the navaid designator is followed by the bearing from the aid (three digits, with leading zeros) and the distance in nautical miles from the aid (three digits, with leading zeros), for example 'DUB180040'. Coordinates and bearing/distance examples are from Doc 4444.

Waypoints precede and follow (separated by a space) each coded designator of an airway, or segment of an airway. The coded designators of airways are comprised of between two and seven characters, for example 'L620' and 'UL620' – airways prefixed with a 'U' being in upper airspace (usually above the equivalent airway in lower airspace).

In summary, the filed route consists of the following parts, with however many points and airways needed to reach the destination:

$$\left(\begin{array}{c} \text{Speed \&} \\ \text{Level} \end{array} \right) \quad \left(\begin{array}{c} \text{SID* ▶ Point1 ▶ AirwayA ▶ Point2 ▶ AirwayB ▶ Point3 ▶ STAR*} \end{array} \right)$$
* Or direct route to/from airways

There is no need to file intermediate waypoints along the same airway if no changes are planned at these points. So for example, 'Point1 ▶ AirwayA ▶ Point2 ▶ AirwayA ▶ Point3 ▶ AirwayA ▶ Point4' is simplified to 'Point1 ▶ AirwayA ▶ Point4' as there is no change in airway between 'Point1' and 'Point4'. IFPS will automatically simplify such routes.

Any change to the cruising speed, cruising level or flight rules is specified at the waypoint (however expressed) where the change is planned. Speed changes are only specified when a 5 per cent true air speed or 0.01 Mach (or more) alteration is planned, whereas all level changes are declared. Where a change is planned, the waypoint is followed by an oblique stroke and then the new cruising speed and cruising level expressed in the same manner as the initial cruising speed and initial cruising level at the start of item 15. Taking the example flight plan, 'POBIX/N0414F230', a change in both speed and level is planned at the 'POBIX' waypoint. Whether cruising speed and/or cruising level are changed, *both* are given at each change. Cruise climb is specified differently. Where a cruise climb is planned, the waypoint is prefixed with 'C/' and followed by an oblique stroke with the speed to be maintained, and the two levels to be occupied (the speed and levels are expressed in the usual manner). For example, if permitted at this waypoint, a cruise climb from FL 330 to FL 370 could be filed as 'C/POBIX/N0450F330F370'. Alternatively, the higher of these two levels (FL 370) could be substituted with 'PLUS', giving 'C/POBIX/N0450F330PLUS'. The point at which a change in the flight rules is planned is labelled with 'VFR' (if changing from IFR to VFR) or 'IFR' (*vice versa*), with a space used to separate the new flight rules from the waypoint or the waypoint/speed/level combination (ICAO, 2001).

Having outlined the components that make up a filed route, focus is now placed on the route chosen for the example flight plan. As discussed earlier, flight planning software and CFMU's CIA and CHMI applications have direct access to sources of aeronautical information (as well as other pertinent information), however, the use of the underlying publications of this aeronautical information (introduced earlier) is now examined to illustrate how the example route was built.

The RAD was consulted to determine the route restrictions in the countries affected by this city-pair (UK, Belgium and Germany RAD Annexes), as were Appendix 3 'City-pair Level Capping' (EUROCONTROL, 2007g) and Appendix 4 'DCT Limits' (EUROCONTROL, 2007h). The airway designated UL607 is a suitably direct route to use between London Heathrow and Frankfurt (refer to Figure 2.3), however the following restrictions apply to UL607 or the city pair:

The UK and Ireland SRD was referred to in order to construct the UK portion of the route (Table 2.3). The SRD offers preferred routings, and, in the case of the example route from Heathrow, suitable SID, routing and connecting point with UL607 (NATS, 2007).

Table 2.2 Relevant Route Availability Document restrictions

ID (source)	Airway segment or city pair*	Restriction*
ED**2044 (Annex ED)	UL607 MATUG – TGO	Not available for traffic Dest. **EDDF**/DS/FE/FM/FV/RY /SB, ETAR/OR/OU
EB**4002 (Appx. 3)	**EDDF**/DG/DK/DL/GS/KZ/LA/LE/ LI/LM /LN/LP/LS/LV/LW, ETAR/ ID/OU, ELLX **to/from London Group**	Not above FL 330
EG**4052 (Appx. 3)	**London Group to/from German Group**	Not above FL 330

* Author's emphasis in bold.

Source: EUROCONTROL, 2007f; EUROCONTROL, 2007g.

Table 2.3   Relevant UK and Ireland Standard Route Document restrictions

ID	Airway segment	Restriction
Note 117	(ADEP: EGLL / SID: DVR / Route: UL9 / Exit: KONAN) UL9 KONAN UL607	FL 250 is not available as a cruising level on UL 607 east of KONAN

Source: NATS, 2007.

In addition, the en-route ('ENR') parts of the AIPs of the countries affected by this city-pair were consulted (CAA, 2007a; Belgocontrol, 2007a; Belgocontrol, 2007b). Departure and arrival procedures at the two airports were checked in the UK and German AIP aerodrome ('AD') parts – more later.

Table 2.4   Relevant Aeronautical Information Publication restrictions

ID (source)	Remarks*	Restriction
5.4.2 (EG ENR 1-9-2)	Where the total sector length (including any portion outside the London/Scottish UIR/FIR[a]) exceeds 220 NM operators are to file a requested flight level for the entire route at FL 250 or above unless prior approval has been given by the UK FMP.	Above or at FL 250
1.1 (EB ENR 3-2-1)	FL 250 is not available for traffic overflying the Brussels UIR.	Above FL 250
1.1 (EB ENR 3-2-1)	FL 330 is the maximum plannable cruising FL within the Brussels UIR for traffic originating **from EDDF**, EDDK, EDDL and ELLX with **destination London TMA[b] and vice versa**.	Not above FL 330
UL607 (EB ENR 3-3-19)	FL 270 not available for flight planning purposes between KONAN and MATUG.	FL 270 is not available

*   Author's emphasis in bold.
a   Upper Flight Information Region/Flight Information Region.
b   Terminal Manoeuvring Area.

Source: CAA, 2007a; Belgocontrol, 2007a; Belgocontrol, 2007b.

As Tables 2.2 to 2.4 illustrate, there are numerous restrictions to consider since the preferred route[28] passes through busy sectors of airspace. In summary, to use UL607 the filed route must be above FL 250, but not above FL 330, excluding

---

28   UL607 offers a route between north-west Europe (Ireland) and north-east Africa (Egypt), via central Europe.

FL 270. Although UL607 can be joined at the KONAN waypoint, Frankfurt cannot be reached beyond the MATUG waypoint.

Returning to the example filed route with these restrictions in mind, 'DVR6J DVR' specifies the Standard Instrument Departure to the connecting point on the first airway. DVR 6J (filed without the space) is one of many SIDs published in the UK AIP under London Heathrow, in the aerodrome part (CAA, 2007d). The departure runway (if more than one runway is available) and direction used will generally vary due to local noise agreements and wind direction – refer to Chapter 5 for a discussion on noise limitations. A suitable SID has been selected for the example route – take-off from runway 09R to the west (Heathrow has a 'westerly preference'), turn right and establish on the Detling ('DET') navaid, crossing set distances from Detling at minimum altitudes (for example 'DET D20 5000' – 5000 ft or above when 20 NM from Detling), to the Dover ('DVR') navaid. DVR is on airway UL9 and shortly after this point the top of climb will be reached, with the aircraft cruising at 444 knots at FL 330. UL9 continues to KONAN (where this airway ends) at which point UL607 is joined.

As discussed earlier, there is no need to file intermediate waypoints along the same airway, hence waypoints KOK, FERDI, BUPAL and REMBA (between KONAN and SPI) are not filed. At the Sprimont ('SPI') navaid, airway UT180 is transferred to, however UT180 is a short airway (only 31 NM) and at DITEL joins airway T180. At waypoint POBIX, the start of descent is planned with the change in speed and flight level declared. The intermediate waypoints BENAK and AKIGO are ignored in the filed route as T180 continues to OSMAX, the starting point of the intended STAR at Frankfurt.

OSMAX 2E (again filed without the space) is one of the published Frankfurt Main STARs available in the aerodrome part of the German AIP (DFS, 2007). This STAR consists of waypoints OSMAX, EPINO, LAGES and REDLI with specified speeds and altitudes at each point, enabling in this case, a planned landing on runway 07L.

IFPS checks the route filed in item 15, for example, that points on the planned airways exist in the environment database; automatically inserting the correct waypoint if the route is found to start with an airway designator (*vice versa* at the end of the route); and verifying speed and levels are compatible with the aircraft performance of the aircraft type in item 9 (EUROCONTROL, 2007b).

Filed Flight Plan: Item 16 (Destination Aerodrome, Total Estimated Elapsed Time and Alternate Aerodrome)

```
-EDDF0057 EDFH
```

As with item 13, the first part of item 16 consists of eight characters. The destination aerodrome is the four character ICAO location indicator (in this case, 'EDDF', Frankfurt Main), and is the third unchangeable key field of the flight plan. If no location indicator exists then the code 'ZZZZ' is used, with the name of the aerodrome included in item 18 using the DEST sub-field. The total estimated elapsed time (EET) of the flight is the time taken from take-off to arrival over the

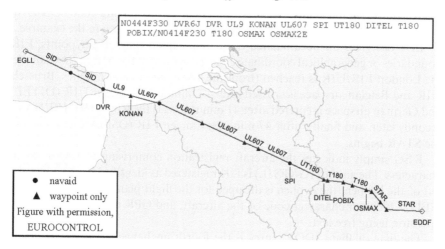

N0444F330 DVR6J DVR UL9 KONAN UL607 SPI UT180 DITEL T180
POBIX/N0414F230 T180 OSMAX OSMAX2E

Figure 2.3   London Heathrow (EGLL) to Frankfurt Main (EDDF) flight plan

designated point from where the Instrument Approach Procedure begins, or to arrival at the destination aerodrome itself, if no navigational aid is associated with it (CAA, 2006d). There is no gap between the location indicator and the EET. The EET consisting of four digits that represent the expected flight time and not the time of arrival – in the example '0057' corresponds to zero hours and 57 minutes. In instances where the flight plan is filed when the aircraft is airborne (item 13 is 'AFIL'), the EET is an estimate from the first point given in the filed route.

Up to two alternate destination aerodromes can be listed in the second part of item 16, required in case unexpected conditions at the intended destination (such as adverse weather) result in a diversion. These alternative destinations are aerodromes that can handle the aircraft type as well as in range of the planned fuel load. The alternative destination in the example flight plan, 'EDFH' is the four character ICAO location indicator for Frankfurt Hahn. If no location indicator exists then, as might be expected, the code 'ZZZZ' is used, with the name of the alternative aerodrome included in the sub-field ALTN, in item 18. Note that although two alternate destinations can be processed by IFPS, only one coded as 'ZZZZ' can be accepted.

Filed Flight Plan: Item 18 (Other Information)

-EET/EGTT0008 EBUR0017 EDVV0041 EDUU0041 EDGG0047 REG/GGGJL
SEL/MITJ OPR/TANNER AIRWAYS DOF/070504 RMK/TCAS)

Item 18 consists of other information relevant to the IFR/GAT flight presented as recognised sub-fields, some of which have already been mentioned in the description of other flight plan items. Each sub-field is identified by a key word followed by an oblique stroke, and the related information (note some are not used in RPLs). IFPS accepts over 20 sub-fields, cross-referencing with previous items where additional information is expected (such as the missing code 'ZZZZ' in items 9, 13 and 16), however this section focuses on those used in the example.

EET describes the accumulated estimated elapsed times at waypoints, FIR boundaries or geographical coordinates,[29] as prescribed in the AIP. In the example, the London FIR/UIR is reached (from the TMA) after eight minutes, Brussels UIR and Belgian airspace is crossed after 17 minutes, Hannover UIR (DITEL) and German airspace is entered after 41 minutes, as is the Rhein UIR (BENAK) seconds later, and finally, after 47 minutes, Langen FIR (OSMAX 2E) – where the STAR begins.

REG simply indicates the aircraft registration comprising of two to seven characters. The aircraft (G-GGJL) is UK registered as identified by the 'G' to the left of the hyphen (the hyphen is dropped for the flight plan). SEL identifies the 'SELCAL'[30] four character code of the aircraft, and OPR indicates the aircraft operator, using free text.

The date of flight (DOF) entry is the fourth unchangeable key field in the flight plan. Although it is recommended to always include the date of flight (formatted as year, month and day with two digits for each – '070504' for the fourth of May 2007), if omitted then the planned flight must be within the next 24 hours. For a flight with an EOBT more than 24 hours away (up to five days in advance) the date of flight is required – and is mandatory for flights in the IFPS Zone (CAA, 2006d).

The RMK sub-field enables information required by an ATS authority, or by the pilot-in-command, to be filed as free text remarks. There is no limit to the number of characters used in the remarks, other than the message as a whole has a maximum of 2100 characters (including header). Multiple remarks can be filed, though IFPS will combine into one sub-field. The example remark concerns TCAS (Traffic Alert and Collision Avoidance System) provision on board.

Filed Flight Plan: Item 19 (Supplementary Information)

Observant readers will have noticed that the example flight plan does not contain any supplementary information after item 18. However, just because the information contained in item 19 (such as the name of the pilot) is not normally filed, it is known by the aircraft operator. Indeed item 19 on the ICAO model flight plan form is captioned with 'SUPPLEMENTARY INFORMATION (NOT TO BE TRANSMITTED IN FPL MESSAGES)'. It is the responsibility of the aircraft operator to make sure that this information is available if an ATS Unit should request it. IFPS will accept, process and store item 19 information if filed.

Item 19 contains up to nine sub-fields of information: the total fuel endurance for the flight in hours and minutes ('E'); the total number of passengers and crew on board, ('P'), if the total is not known at the time of filing then 'TBN' (to be

---

29 As expressed in the item 15 route.

30 'Selective Calling': aircraft-specific radio addressing equipment that enables ground to contact the aircrew even when the audio level is turned down in the cockpit.

notified) is inserted; the emergency radio capability ('R') with up to three types and frequencies (such as 'U' for UHF on frequency 243.0 MHz); the survival equipment capability ('S') with up to four types (such as 'M' for maritime survival equipment); the life jacket capability ('J') with up to four standards (such as 'L' for jackets with lighting); the dinghy capability ('D') in terms of the number, their capacity, whether they are covered and their colour; the aircraft's livery and significant markings ('A'); any other survival equipment on board and any other useful comments concerning the flight ('N'); and finally, the name of the pilot-in-command ('C'). A completed item 19 for the example flight plan follows (with no remarks or other survival equipment is on board).

```
-E/0125 P/136 R/VE S/M J/LF D/4 160 C ORANGE
A/BLACK WITH AMBER TAIL GGGJL C/BROUGH
```

Note that if a written or printed ICAO flight plan form was filed (for a VFR flight), equipment that is unavailable is crossed out on the form.

2.4   Slot Allocation and Messaging

Having discussed the information used to plan IFR flight plans, the contents of the submitted plan (including aircraft type, departure and destination airports, EOBT and route) and distribution to CFMU, this section considers how flight plans enter the tactical flow management phase with particular attention given to slot allocation.

2.4.1 Enhanced Tactical Flow Management System (ETFMS)

When CFMU assumed responsibility for the European ATFM service in 1996, tactical and pre-tactical operations were supported by the TACT system. This system had two key functions: the calculation of traffic demand in every airspace sector within the CFMU Area, using planned information (filed flight plans) received from IFPS; and the calculation, allocation and distribution of slot lists by the Computer Assisted Slot Allocation (CASA) system – see next section. These functions allowed a comparison to be made between planned traffic demand and declared departure airport, arrival airport and en-route sector capacities. Where capacity was exceeded, FMP and CFMU flow management controllers would be notified, in order for a decision to be made as to whether flow restrictions were necessary – whereby the most constraining airport, sector or ACC, determines the volume of traffic which can be handled for a given flow. Flow restrictions, known as 'ATFM regulations' or simply 'regulations', are still implemented by CFMU and work by delaying affected aircraft at the departure airport rather than airborne. For example in 2006, 1246 sectors and airports were protected with over 100 ATFM regulations per day – though to put this in context, 35 sectors were responsible of 40 per cent of the en-route delays and just 37 airports accounted for 90 per cent of the aerodrome delays (of which London Heathrow and

Frankfurt were the two worst, if weather is included as a cause). Zagreb, Zurich and Madrid were the most constraining ACCs – in 2006 as a whole, regulations delayed around 10 per cent of flights, compared with around 17 per cent in 2001 (EUROCONTROL, 2007i) – see Chapter 4 for further discussion on this subject. Delaying flights (on the ground) intending to use regulated sectors or airports, effectively moves excess traffic out of the overload time period (Breivik, 2003).

In February 2002, the TACT system was replaced by the Enhanced Tactical Flow Management System (ETFMS). ETFMS has additional functionality in that planned information[31] initially in the system is updated using surveillance (radar) data from ANSPs and position reporting data from aircraft operators, combined with meteorological data (wind speed and direction) every six hours. The availability of accurate real-time (or near real-time) information permits ETFMS to recalculate the 4D profile of flights, allowing traffic demand to be understood more precisely by flow management controllers. For example, whereas the TACT system assumed an aircraft was airborne as expected, ETFMS assumes an aircraft is not airborne until information confirming the situation is received, reducing unnecessary flow management actions.

Flight plan messages are copied from IFPS to ETFMS in ATS Data Exchange Presentation (ADEXP) format. This is a text format rather than a protocol, and though readable, is intended for computer to computer ATS message exchange between aircraft operators (AOs), Air Traffic Service Units (ATSUs), Air Traffic Control Units (ATCUs) and CFMU – examples of ADEXP-formatted message types, in addition to filed flight plans, include the Slot Allocation Message (SAM) and the Slot Revision Message (SRM) – Figure 2.4 provides an overview. Each field of information conveyed by ADEXP messages is delimited with a hyphen ('Start-of-Field' character) followed by a specific keyword (note that the maximum message length is approximately 10 000 characters). For example '-ADEP EGLL' would be used to report London Heathrow in the aerodrome of departure information field (EUROCONTROL, 2001b).

2.4.2   Computer Assisted Slot Allocation (CASA)

Filing a flight plan to IFPS represents a request for an ATFM departure slot. The Estimated Off-Block Time (EOBT) declares the true time the aircraft operator expects the aircraft to be ready to depart. Whether the aircraft is permitted to depart at this time depends on the effect of any flow restrictions placed on the aerodromes and the airspace through which the route is planned. A flight affected by an ATFM regulation receives an 'ATFM slot' (also known as a 'departure slot', 'CTOT' – see later, or just 'slot') from CFMU. (Airport slots are discussed in Chapter 4.)

The process for calculating slots is automatic and determined by the Computer Assisted Slot Allocation (CASA) system – a function within ETFMS that operates under the 'First Planned – First Served' principle. CASA initially allocates an

---

31   ETFMS has flight data for the following 48 hours (RPLs initially, and FPLs when available).

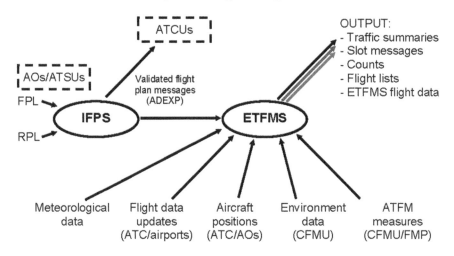

Figure 2.4   Overview of a flight plan message in ETFMS

Estimated Take-Off Time (ETOT) for each flight based on the EOBT plus the taxi-time at the departure aerodrome. This enables each flight to be given an Estimated Time Over (ETO) for the point of entry at each sector used in the planned route, and so be included in these sectors' slot lists. Airports and sectors have declared capacities, and although flow managers will attempt to match capacity with demand (for example by coordinating a temporary increase of capacity of a sector), when the number of aircraft in a slot list exceeds capacity, a regulation is activated.

When a regulation is activated in ETFMS, the key restriction parameters that determine which flights require slots are location, start time, end time and entering flow rate. CASA extracts the planned flights that will enter the regulated airspace and, based on each flight's Estimated Time Over (ETO), sequences them using the regulated flow rate in their arrival order had no restriction been activated. If the ETO slot is free the flight receives no delay, if the ETO slot has already been pre-allocated to another flight then the slot will be allocated to whichever flight planned to overfly the congested location first. (Note that 'First Planned' refers to the ETO based on the flight plan rather than which flight plan was filed first). So the flight not allocated this slot will get the next available one and hence receive a delay amounting to the difference in slot times. This process often leads to a chain reaction of slot changes as new flights enter the slot list. From this, the calculated take-off time (CTOT) is determined – that is, the departure slot. When a flight is subject to multiple regulations, the delay of the most penalising regulation takes precedence and is forced into all other regulations, rather than the ETO. At Slot Issue Time (SIT), at the earliest two hours before EOBT, the Slot Allocation Message (SAM) is distributed to the aircraft operator and the tower at the departure aerodrome informing both of the allocated departure slot (EUROCONTROL, 2006g).

2.4.3   Slot Management

A range of measures are in use today as the European ATM system is becoming more flexible in managing flows. Examples include dynamically adjusting ATM capacity (through measures such as level capping and configuration change) and re-routing (EUROCONTROL, 2007j). However the use of ATFM regulations by delaying aircraft on the ground to prevent over-delivery (receiving more traffic than the coordinated capacity) is a daily reality for aircraft operators. In this section the example Heathrow to Frankfurt flight plan will be referred to, note that all times are UTC.

The flight plan for flight 'GST123' was filed to IFPS and having received an Acknowledgement message (ACK), is in the system. The EOBT is 1500, so with a taxi-time of 15 minutes at Heathrow, the Estimated Take-Off Time is 1515. Considering this is a flight between two congested airports, receiving a regulation is not unusual – the daily ATFM Notification Message (ANM) can be consulted to see which regulations are likely to affect the flight. Staff in the operations department at 'Tanner Airways' use flight planning software with access to both IFPS and ETFMS (as messages are sent and received to/from both systems). If a Slot Allocation Message (SAM) is not received from ETFMS then this flight is not subject to any regulation and can continue with the intended off-block time (note that it is possible to receive a SAM very close to the EOBT). Figure 2.5 shows an example SAM (ADEXP format) that applies to this flight, issuing an ATFM slot of 1540. The majority of the field keywords are familiar to those discussed earlier in the context of a filed flight plan, however two are of interest. 'REGUL' identifies the location, part of the day and date of the regulation[32] affecting the flight (if several REGUL fields are listed, the first one is the most penalising regulation) – in this case Frankfurt ('EDDF'), in the afternoon ('A') on the fourth of the month ('04'). 'REGCAUSE' provides ICAO and IATA codes for the cause of the most penalising regulation – in this case weather at the destination (ICAO 'WA' and IATA '84') (EUROCONTROL, 2006g).

```
-TITLE SAM -ARCID GST123 -IFPLID AA12345678 -ADEP EGLL
-ADES EDDF -EOBD 070504 -EOBT 1500 -CTOT 1540 -REGUL
EDDFA04 -TAXITIME 0015 -REGCAUSE WA 84
```

Figure 2.5   Example of a Slot Allocation Message

With a CTOT of 1540, this flight is being delayed at Heathrow by 25 minutes. To comply with this slot, the Frankfurt-bound flight must take off within a -5 to +10 minute slot window, that is, between 1535 and 1550 (to take into account the ATC departure sequence). If compliance with this slot is not possible, although

---

32   'EDDFA04' is the regulation identifier and can be viewed in the current ANM.

the new EOBT is known, then the flight plan requires updating at the earliest possible time by sending either a Delay message (DLA) or Modification message (CHG) to IFPS. If the new EOBT is not known then a Slot Missed Message (SMM) should be sent to ETFMS, resulting in a Flight Suspension Message (FLS) being returned – the flight remaining suspended until a new EOBT is provided by a DLA message. There can be a reluctance for aircraft operators to 'hand back' a slot if there is a ground-based problem, on the assumption that the fault can be rectified in time for the original slot to be made, however failing to use or release the allocated slot results in the slot being lost from the network – further implementation of the Collaborative Decision Making (CDM) process may change this conduct.

The level of delay and the importance of the connectivity of the flight are just two variables that may influence the effort expended within an operations department in attempting to improve the slot situation (a central theme of Chapter 4). Some activities are deemed unacceptable behaviour by the CFMU Service Agreement – such as 'filing a flight plan for a flight whilst another flight plan for the same flight is still active' (EUROCONTROL, 2004b). A 'true revision process' takes place after the SAM has been issued, which automatically attempts to improve the allocated slot. The default status with ETFMS for all flights is that of 'Ready For (direct) Improvement' (RFI), so when a slot improvement is possible, a Slot Revision Message (SRM) is automatically received by the AO. By sending a SIP Wanted Message (SWM) to ETFMS changes the default status of a flight, so that a Slot Improvement Proposal Message (SIP) will be received instead. However, after reception of a SIP, the AO has to decide whether to accept or reject the proposed slot change by returning either a Slot Improvement Proposal Acceptance or Rejection Message (SPA or SRJ) before the proposal is timed-out and the slot released for use elsewhere. To return the default status back to RFI, a RFI message is sent to ETFMS.

Another way of attempting to reduce delay is to consider re-routing the flight or using an alternative flight level to avoid a regulation (EUROCONTROL, 2006g). Daily ATFM Information Messages (AIM) provide instructions and information about current ATFM measures. In addition to individual AIMs, a summary 'Network News' AIM with excerpts from the ATFM Daily Plan (ADP) may offer level-capping and re-routing suggestions. Whether such information is consulted, an AO can submit a new route by using either a CHG, or Cancellation Message (CNL) and re-filing using the Replacement Flight Plan Procedure (RFP). Alternatively for flights entirely within the IFPS Zone, the Aircraft Operator 'What-If' Re-route (AOWIR) function can be queried using the CIA (CFMU Internet Application) and CHMI (CFMU Human Machine Interface) applications – see earlier section. This function allows the Route Catalogue to be consulted and available re-routings assessed – 41 000 routes for 18 691 city pairs in 2004 (EUROCONTROL, 2005a). Although there are limitations to this function, if an acceptable re-routing is found and selected, the AO can either initiate a CHG message with no further action required, or initiate a CNL message in order to receive a Re-routing Notification Message (RRN) from ETFMS, and then file a new FPL with a route consistent with the one provided by the RRN. Hopefully these actions will result in an improved CTOT, though the route chosen may result in increased fuel burn for the flight (see Chapter 4 for a discussion of the cost trade-offs associated with this).

Other options available to aircraft operators involve making a variety of requests. If the aircraft doors are closed and the flight is ready to depart at a time between EOBT minus 30 minutes and CTOT, the AO may ask the tower to send a ready (REA) message to ETFMS (only the local ATS Unit is able to send this message type). If a slot improvement is possible, the AO will receive a SRM. Slot swapping trials have taken place at a small number of airports, whereby CFMU will attempt to swap the slots of two flights subject to the same most penalising regulation, following a request from the AO, or FMP. A third form of request involves calling the Central Flow Helpdesk in situations where the AO is without access to CFMU systems, when a problem is not solved by the use of ATFM messages, or when the delay to an aircraft is significantly above average (EUROCONTROL, 2006g).

Reception of a Slot Requirement Cancellation (SLC) message informs the AO that the affected flight is no longer subject to an ATFM slot and may depart at the intended time.

2.5   Conclusions

The ICAO flight plan has been around for a long time, and although the process described in this chapter clearly works, there are shortcomings with the contents and format. For example item 18 (other information) has become an ever expanding list of important information as there is no other field in the flight plan to record it. In comparison, the operational flight plan generated by flight planning software and used internally by aircraft operators contains much more detailed information about all aspects of the flight. In terms of the format, it is telling that once FPLs have been processed by IFPS, they are reformatted and distributed as machine-readable ADEXP messages.

A number of studies are currently examining ways of improving the present flight plan, such as the ICAO Flight Plan Study Group and the EUROCONTROL Specification for the Initial Flight Plan. Future developments may include the 'Flight Object', with a single reference for all systems for information relating to a flight; and the integration of 'airport slots' (strategic) at coordinated airports with 'ATFM slots' (tactical), thus making a step towards tighter coordination of these currently somewhat disparate processes.

Chapter 3

Understanding En-Route Sector Capacity in Europe

Arnab Majumdar
Imperial College London

3.1 Introduction

Air travel in Europe continues to grow at a rapid rate. For example, a real annual growth of 7.1 per cent was recorded in the period 1985–1990, greatly exceeding a prediction of 2.4 per cent annual growth (EUROCONTROL, 1987). In 2005, the growth of controlled air traffic over Europe was 3.9 per cent and further examination reveals that growth in Eastern and Central Europe exceeded this figure (EUROCONTROL, 2006a). Unfortunately, this growth has not been matched by availability of capacity. Europe suffers from both runway and en-route airspace capacity limitations (Arthur D. Little Limited, 2000). The latter is heavily dependent on controller workload, i.e. the physical and mental work that controllers must undertake to safely conduct air traffic under their jurisdiction through en-route airspace (Majumdar and Ochieng, 2002). In 2006, EUROCONTROL's long-term forecast (EUROCONTROL, 2006b) gives an indication of the forecast traffic to be expected over Europe by 2025. The forecast considers four main scenarios, as shown in Table 3.1.

These four scenarios indicate that by 2025, air traffic over Europe will increase in the range from 1.7 to 2.1 times 2005 traffic levels, with the strongest growth seen in Eastern Europe, the Baltic States and Turkey. Considering the highest growth scenario, in 2020, traffic demand is forecast to be around twice as high as 2005 levels (SESAR Consortium, 2006b), with the consequence that there will be approximately 18 million fights in 2020. The European ATM system will struggle to accommodate such growth and there are predictions that, in 2020, around 60 airports will be congested, while the top 20 airports may also be saturated for 8–10 hours a day.

The consequences of on-going growth on the current European ATM system are reflected, in part, in delays and flight inefficiencies. The fragmented nature of the European ATM system also hinders its ability to cope with the growth in air traffic. The costs of fragmentation to the European en-route ATM/CNS (Communication, Navigation and Surveillance) system have been estimated to be in the range €880–1440 million, accounting for 20–30 per cent of annual en-route costs, and this contributes to the performance gap between Europe and the US

Table 3.1   EUROCONTROL forecast traffic scenarios

Scenario	Description
Scenario A	Greater globalisation and rapid economic growth, with free trade and open skies agreements encouraging growth in flights at the fastest rate.
Scenario B	'Business as usual', with moderate economic growth and no significant change from the *status quo* and current trends (Note: European Union expansion is at its fastest in this scenario).
Scenario C	Strong economies and growth, but with strong government regulation to address growing environmental issues. As a result, noise and emission costs are higher, which encourages a move to larger aircraft and more hub-and-spoke operations. Trade and air traffic liberalisation is more limited.
Scenario D	Greater regionalisation and weaker economies leading to increased tensions between regions, with knock-on effects limiting growth in trade and tourism. Consequently, there would be a shift towards increased short haul traffic. Security costs increase further beyond 2010, with the price of fuel being at its highest in this scenario, reaching close to 40% of the airline operating costs by 2020 and beyond.

in ATM (EUROCONTROL, 2006h; see also Chapter 7). The main components of this fragmentation cost appear to be the:

• non-optimal economic size of many Area Control Centres (ACCs);
• duplication of bespoke ATM systems;
• duplication of associated support.

The way in which the capacity shortfall is handled through the imposition of ATFM regulations and departure slots was discussed in Chapter 2. The costs of the associated delays will be discussed in Chapter 4, and we will return to a discussion of growth in Chapter 6.

The objective of this chapter is to identify and analyse the methodologies employed to estimate the capacity of en-route airspace, highlighting the intricacies involved in this process, and thus helping towards a better understanding of the issues involved.

3.2   Fundamentals of En-Route Airspace Capacity Estimation

A number of initiatives over the years have addressed the issue of capacity in Europe, notably the EUROCONTROL 'ATM 2000+ Strategy' programme

(EUROCONTROL, 1998; this being continuously updated and refined) and the European Air Traffic Control Harmonisation and Integration Programme (EATCHIP[1]). Two of the key elements of these initiatives are new technologies and control procedures. A vital point to note here is that the success of such initiatives depends to a large extent upon a reliable definition, measure and assessment of en-route airspace capacity. A major hurdle in this area is that airspace capacity in high traffic density areas, such as Europe, is not determined solely by spatial-geometric constraints dependent on aircraft performance criteria. Rather, as we touched upon in Chapter 1, it is the controller workload that is the key driver and this brings with it associated difficulties in the measurement of this workload (European Commission, 2002).

ANSPs (Air Navigation Services Providers, see Chapter 1) commonly use 'fast-time' simulation (FTS) techniques, i.e. computer modelling of controller workload, to estimate en-route capacity. FTS allows greater flexibility for capacity estimation, facilitating investigation of the impacts of a wider range of proposals. The outputs of the simulations are then usually post-processed to formulate a relationship between the number of aircraft entering the sector and the workload associated with controlling the air traffic in the sector over a given period of time. However, there has been dissatisfaction with this methodology since research increasingly indicates that the amount of air traffic alone is inadequate to explain the workload experienced by a controller (Mogford et al., 1995; Laudeman et al., 1998). There is an increasing awareness that the performance of controllers is affected by the complexity of the traffic they handle, the structure and geometry of the airspace they control and interaction between these two. Recent research has led some ANSPs to derive a functional relationship between a number of complexity factors, i.e. air traffic and sector factors, and controller workload, at an aggregate level.

Such capacity estimates from FTS alone are insufficient, lacking the human elements such as controller judgement and thinking. Therefore, FTS is often followed by 'real-time' simulation (RTS), which involves building an operational environment, complete with the technologies to be tested, as well as pseudo-pilots, i.e. pilots situated in an adjoining room to the control room, communicating with controllers. The main drawback of RTS is that it is costly, requiring personnel training, infrastructure, significant simulation time and a high-fidelity simulation of the operating environment (Magill, 1997). However, currently, RTS is essential in estimating the impact of new technologies and procedures on controller workload and capacity.

En-route airspace capacity can be defined in purely spatial criteria as the maximum number of aircraft through any given geometrical airspace for a given time period, based upon the spatial control constraints which govern the internationally specified separation between any two aircraft given their performance characteristics (EUROCONTROL, 1991). As air traffic increases, the

---

experience in high-density traffic areas, suggests that a safer measure of capacity should be based on air traffic controller workload (Majumdar and Polak, 2001). The introduction of controller workload poses problems in capacity estimation as it is a confusing term, with a multitude of definitions and models in the literature, and its measurement is not uniform (Jorna, 1991). Workload is a construct, i.e. a process or experience that cannot be seen directly, but must be inferred from what can be seen or measured. Research, theory, models and definitions of workload are inter-related and there are numerous reviews of workload and its measurement (Gawron et al., 1989). The methods for measuring controller workload are usually categorised in three groups: subjective, objective and physiological (Hopkin, 1995). Subjective self-assessment of controllers is carried out either by instantaneous or non-instantaneous techniques, while objective methods involve direct observation and recording of the actions of controllers, usually by other controllers, or ATC system experts. Physiological methods, e.g. heart rate measurements, are not used to the same extent as the other methods due to the difficulties posed in their implementation on the operational control room, or during RTS.

In Europe, the capacity of an ATC sector is defined as the 'maximum number of aircraft that can enter the sector in a specified period', while still permitting an 'acceptable level of controller workload' (Sillard et al., 2000). The workload is usually assessed based on task-time definitions obtained from a detailed, non-intrusive, objective record of the controller's actions, aided by controller verification. The task times are then used to determine threshold controller loadings in units of minutes/hour, as recorded by the FTS models used (EUROCONTROL 1999a; Stamp, 1992).

So, although the focus of conventional capacity assessment has been on aircraft flow rates, recent research has indicated that the workload experienced by controllers is affected by a complex interaction of a number of factors (Mogford et al., 1995). These include the situation in the airspace (i.e. features of both the air traffic and the sector), the state of the equipment (i.e. the design, reliability and accuracy of equipment in the control room and in the aircraft), and the state of the controller (including age, experience, and decision making strategies). The effect of these factors on airspace capacity must be understood if realistic and successful strategies for increasing capacity are to be implemented. Increasing research effort is being directed at determining the factors associated with the features of air traffic and sectors (i.e. ATC complexity factors), and quantifying the link between controller workload and a number of such complexity factors. Hilburn (2004) provides a good review of such studies, and we will return to this later.

3.2.1   *En-route Airspace Capacity Estimation based on ATC Workload*

Historically, en-route airspace capacities in some European nations have been determined based on operational experience. This involved investigation of how much traffic a particular sector could handle based upon controller experience of that sector. Typically, an observer would count the traffic into the sector over a given period of time, and follow this by conducting controller interviews to ascertain the controllers' perception of the traffic load. Based on these two factors, a capacity figure would be determined as the capacity to be declared to flow control authorities (flow control was discussed in Chapter 2). Any subsequent information would later refine this declared capacity estimate. Such a method means that the capacity is estimated empirically based upon judgements gleaned from the operational expertise of controllers. Every season, operational evaluations would be undertaken to determine the capability of controllers to cope with a reference scenario.

The situation in Europe has evolved since the mid-1990s. The basic process for the Master ATM European Validation Plan (MAEVA) has been described by Revuelta (2000). MAEVA originally provided the European Commission with a consolidated, common and consistent validation strategy for the future European ATM Programme (EATMP) including projects funded under the Fifth Framework Programme (FP5), and is now used in EATM. MAEVA supports the specification of the required validation environment for individual validation exercises and the ATM components to be integrated. The main ATC/ATM providers in Europe now follow the MAEVA process, and one of the earliest stages involves the use of FTS which enables new ATC/ATM concepts to be analysed. In addition, capacity aspects due to redesign of airspace and changes in air route structures are also investigated by FTS.

3.2.2   *Methodology to Estimate En-Route Airspace Capacity*

The capacity of a simple system of ATC sectors depends on a number factors (Stamp, 1990), including the physical pattern of routes and airports, the pattern of traffic demand (both geographic and temporal), and ATC routing procedures designed to maximise traffic throughput. The typical procedure adopted in Europe to estimate capacity recognises these factors and consists of a number of key stages, as represented in Figure 3.1 and given below:

- Definition of the physical characteristics of the air traffic system. This is conceptually simple for the current ATC network, for example, the geographical airspace sectors and the number of flight levels in each sector. While these are directly observable, data collection and analysis can be time consuming.
- Definition of the ATC procedures and Standing Agreements for the airspace sectors and their horizontal and vertical boundaries (geographical locations).
- Definition of the air traffic handled through the sectors. This involves definition of the aircraft types and their associated performance characteristics.
- Definition of a set of controller tasks and their timings for the particular Area Control Centre (ACC) being simulated. This step also involves calibration of the tasks for that centre and is based on video and audio recordings of controllers at work, followed by controller questionnaires and interviews.
- FTS modelling which involves systematically varying a number of possible airspace scenarios and traffic parameters and investigating their interactions.

Again, this is a demanding task, especially when future traffic scenarios are considered, as this requires the assessment of future fleet plans of airlines together with consideration of improved aircraft performance.
- Statistical analysis: from the output of the simulation model, and based upon a calibrated set of controller tasks, their timings and frequencies in a sector, statistical analysis is undertaken to derive a functional relationship between airspace capacity and a number of relevant parameters. This requires definition of the rules for assessing the capacity of an individual ATC sector for use in the workload simulation model.

The key assumptions made in the formulation and execution of the procedure above include the following.

- In Western European en-route airspace sectors, the current limiting factor on airspace capacity is controller workload. Even for future ATC systems, controllers will remain an essential part of ATC, though their tasks will be expected to evolve.
- Controller workload can be broken down into a series of physical tasks, for example radio/telephony (RT) communications and flight data management, whose timings can be determined. The cognitive tasks can also be assessed and quantified, especially for the crucial tasks of conflict detection and resolution.
- En-route sectors are usually assumed to be controlled by a control team of a 'planner' or planning controller (PC) and a tactical controller (TC), each 'position' with specified roles and tasks (as introduced in Chapter 1). Their current roles and tasks are such that the TC has a considerably higher workload than the PC and is the bottleneck in European en-route capacity.
- The simulation model chosen realistically reflects the 'real world' airspace environment being simulated.
- The calibrated tasks and their timings apply to all sectors in an ACC.

Such a methodology has the advantage of considerable thoroughness. However, this process is resource intensive, with calibration, in particular, taking considerable time and is prone to a subjectivity bias on the basis of controller interviews. Additional resources would be needed if RTS is also required.

3.2.3   Controller FTS Workload Models

There are two main models for simulating controller workload:

- Re-Organised ATC Mathematical Simulator (RAMS) (EUROCONTROL, 1995);
- Total Airspace Airport Modeller (TAAM) (Odoni et al., 1997).

EUROCONTROL also employs a methodology to assess capacity known as the Capacity Analyser (CAPAN), which incorporates either the European

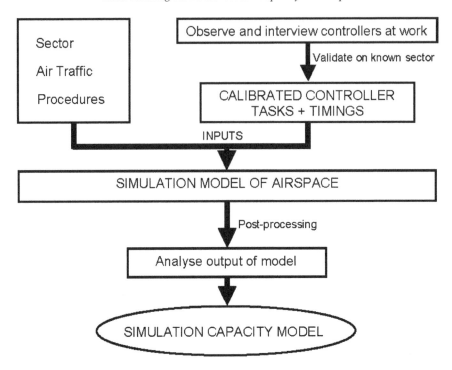

Figure 3.1   Estimation of capacity using a model of controller workload

Source: Adapted from Majumdar and Polak (2001).

Airspace Model (EAM), a predecessor of RAMS, or RAMS itself. In addition, a model of air traffic controller workload based upon the cognitive tasks of a controller was developed by the UK's National Air Traffic Services (NATS), known as 'PUMA' (Performance and Usability Modelling in ATM).

In essence, all the FTS models are discrete (critical) event simulation models. Thus, during the simulation, the model generates a number of defined events in the life-cycle of a simulated flight for each sector crossed. At each event, a number of actions are determined and system recordings made, and these events are used to generate tasks and workloads to control positions based on task definitions and achieved flight profiles. Example events include entry into the first simulated sector, exit from a sector, conflict search and conflict resolution. In that respect, it is important to note that both RAMS and TAAM are based on objective measurements of controller tasks and are better described as taskload models. For the purposes of this chapter, they will still be referred to as controller workload models.

In the US, an analytical model of air traffic controller workload (Sector Design and Analysis Tool – SDAT), not based upon simulation, is also used to provide traffic and other data to estimate sector capacity. SDAT is limited to flight trajectory calculations without consideration of flight conflicts. It uses historical radar track and RT communications data for a sector. A variety of metrics are provided by SDAT that are used as indicators of sector capacity, including: sector density (the number of simultaneous flights within a sector); sector throughput (the number of flights entering/exiting a sector); the percentage of flights climbing, cruising and descending; the time flights are within a sector; and, flight crossings (both their locations and classification – by overtaking, horizontal crossing and vertical crossing). SDAT supplements the FTS conducted by the FAA to estimate sector capacity.

On completion of the FTS, an analysis package examines the resulting profiles of each aircraft and determines a defined number of tasks that were required of the controllers to process the flight. As each controller task has a defined execution time and associated 'position(s)', it is possible to determine the amount of work required to handle a given traffic sample. The models have four main types of control and input data: airspace structure and route network, traffic samples, ATC logic and procedures, and controller task definitions.

When used for capacity assessment, all the data and parameters given above are specified as initial inputs to the model, then simulated, and the results analysed for conformity with the specifications (such as capacity threshold). Thereafter, the only data changing between iterations of the model is the traffic sample.

We will later expand on TAAM and RAMS from an application perspective. Further details on the theory and modelling concepts employed within the models can be found in Sillard et al. (2000). Table 3.2 highlights the main features of the simulation models currently used in capacity estimation.

3.3   Methods of Estimating Airspace Capacity using FTS

An analysis of the methods employed to estimate en-route airspace capacity around the world is presented in a survey by Majumdar et al. (2005). As mentioned, although en-route airspace capacity can be determined from spatial-geometrical constraints, as air traffic increases, controller workload is the main driver. Therefore, the focus of the survey was to analyse en-route sector capacity estimation methodologies in countries with high traffic densities, primarily in Europe, North America and Japan. Table 3.3 gives details of participating ANSPs. The authors captured data from two main sources: existing literature in the public domain and interview-based questionnaires. The latter was aimed at planners and managers of en-route airspace, i.e. those who were intimately involved in the airspace design and sector capacity estimation process, in specific countries. The questionnaires determined:

• whether the ANSP routinely estimated the capacity of new airspace sectors to support airspace (re-)design;
• elaboration of the methodology used by the ANSP to estimate capacity during airspace (re-)design. Particular attention was given to the details of the application of FTS and the link to capacity, including questions on the

Table 3.2   A comparison of air traffic controller workload models

	SDAT	TAAM	RAMS	PUMA
Developer	FAA (US)	Preston Group (Boeing)	EUROCONTROL	NATS (UK)
Type	Analytic	Simulation	Simulation	Simulation
Task data source	Input and output messages of the HOST/SAR* computer system tapes; observer data; facility voice tapes	European ATC reference tasks based on controller observation	European ATC reference tasks based on controller observation	Observational Task Analysis (OTA), including cognitive debrief
Controller	PC and TC	TC	PC and TC	Can adapt for both PC and TC
Taskload differentiation	Good routine tasks; probabilistic conflict resolution; accounts for sector scanning and planning tasks	Good routine tasks; simulates deterministic conflict detection/ resolution; accounts for some scanning and planning tasks; no general monitoring; good coordination tasks	Good routine tasks; simulates deterministic conflict detection/ resolution; accounts for some scanning and planning tasks; no general monitoring; good coordination tasks	Excellent modelling of cognitive aspects of the controllers' tasks; should be able to account for all tasks done by a controller – observable and non-observable
Validation and use	US only, validation in progress	Strong validation of the method in Germany and Switzerland	Together with EAM, validated for Europe's airspace.	Validation in progress
Strengths	Requires little or no manual input from the user – thus no need for manual data collection and no 'guesswork' on the user's part; useful for rapidly determining the workload of controllers in a sector, when compared to full scale fast-time or real-time simulation	Considerable adaptation by DFS in Germany ensures that the method has been widely used for capacity assessment; good conflict detection and resolution.	Widely used and validated throughout Europe for different European airspace sectorisations; a very flexible model and relatively easy to use; teamwork aspect of ATC relatively well modelled; good conflict resolution mechanism based on rules for European airspace	Only model which considers the cognitive processes of the controller; models controllers undertaking more than one task simultaneously; has been used to some extent in the European context for improving real-time and fast-time models

Table 3.2 cont'd

	SDAT	TAAM	RAMS	PUMA
Weaknesses	Largely limited to what can be determined or inferred from the recorded SAR* data, which is only a sub-set of the entire task load of the controller; assumes that no tasks are performed simultaneously and that all task actions occur when the data were recorded by the HOST – not necessarily always the case; limited to historic analysis of sector data, so cannot be used for planning future traffic scenarios	Simulations can take considerable time; workload elements need to be better defined and monitoring tasks need to be better modelled; calibration of the parameters requires effort and the parameters obtained in the model for Germany and Switzerland are not valid elsewhere; cognitive tasks are not accounted for; assumes that two tasks cannot be carried out simultaneously	Does not appropriately account for the cognitive aspects of air traffic control; assumes that two tasks cannot be carried out simultaneously	Cumbersome model – Observational Task Analysis difficult to carry out; any change from base case requires new OTA and real-time simulations; not fully validated for use

* The main data sources are the messages contained in the computer system used for flight data processing in the Air Route Traffic Control Centers (ARTCCs) in the US (known as the HOST system). This system is used to control en-route air traffic. The System Analysis Recording (SAR) service within the National Airspace System of the US, provides magnetic tape recordings which are initiated by specific requests within the ATC subprograms.

Source: adapted from Majumdar and Polak (2001), Sillard et al. (2000).

tools used, the assumptions that underlie the modelling and, importantly, on the post-processing of the simulation outputs;
- the role of RTS in assessing sector capacity, and the balance between RTS and FTS in such assessments;
- the time taken by the various components of the capacity estimation process and matters relating to controller involvement.

Table 3.3   ANSPs in the survey

ANSP	State
Aeropuertos Españoles y Navegación Aérea (AENA)	Spain
Belgocontrol	Belgium
Centre d'Etudes de la Navigation Aérienne (CENA)	France
Deutsche Flugsicherung (DFS)	Germany
Federal Aviation Administration (FAA)	US
Luftfartsverket (LFV)	Sweden
Ministry of Land, Infrastructure and Transport (MLIT)	Japan
National Air Traffic Service (NATS)	United Kingdom
NAV Canada	Canada
Nav Portugal (Navegação Aérea de Portugal)	Portugal
Sistemi Innovativi per il Controllo del Traffico Aereo (SICTA)	Italy
Skyguide	Switzerland
EUROCONTROL	On request from member states

3.3.1   TAAM-based Methods

The major nations using the TAAM simulation model are Germany, Switzerland, France, Canada, Japan and the US. The basic approaches to capacity estimation using TAAM discussed below (and illustrated in Figure 3.2) can be divided into two groups:

- the DFS workload model (in which capacity is estimated using a formula);
- the methodological approach (in which the workload calculation is used in a comparative manner and as one element in estimating capacity).

The DFS workload model approach Although TAAM was not originally designed for en-route controller workload estimation (Sillard et al., 2000), DFS modified it to include en-route sectors. A key element in this was the consideration of the workload of the tactical controller (TC). Based on this, DFS developed a formula to estimate en-route airspace capacity as a function of different drivers of workload, including the number of flights per hour, the average flight time per sector, the number and severity of potential conflicts, coordination (quality and number of the coordination units) and the number of vertical movements.

Table 3.4 shows the workload share of an average ACC sector based on the experience of more than 30 simulations in German ACCs. Based on this, and following discussions with controllers, a measure for the highest acceptable workload was defined at 90 TAAM workload points. The DFS workload model is currently used in Germany and Switzerland.

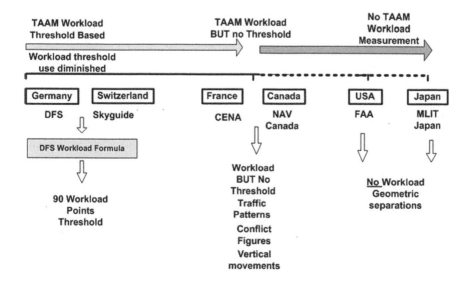

Figure 3.2   Basic approaches to capacity estimation using the TAAM model

Table 3.4   The Workload Share in the DFS Model

Symbol	Sub-workload	Percentage share
WL_1	Movement	50–60
WL_2	Conflict	15–30
WL_3	Coordination	10–15
WL_4	Level change	10–15

Source: private communication, DFS, April 2004.

The overall TAAM workload is calculated by adding the four sub-workloads:

$$\Sigma\, W = WL_1 + WL_2 + WL_3 + WL_4 \qquad\qquad [1]$$

$$\Sigma\, W = \text{Total workload}$$

In order to estimate the capacity, a number of traffic samples (current and future) are simulated using TAAM, for example, with different proportions of aircraft in climb and descent, until the simulations reach the threshold workload value. Therefore, DFS consider not only the number of aircraft but also the characteristics of the air traffic pattern simulated at controller-threshold capacity. This range of capacity (based on the simulation) can be defined as, for example, 48 aircraft with 90 workload points when 80 per cent of traffic is overflying the sector; and 42 aircraft when the majority of flights are departing and arriving traffic (climbs and descents). This is then discussed with the controllers and the board of supervisors for the ACC and one capacity value is then declared to the Central Flow Management Unit (CFMU) in Brussels (as discussed in Chapter 2). The model's strength is that field (operational) data has shown that it is accurate in predicting the range of capacities for different sectors in Germany.

The threshold workload value of 90 workload points was determined from analysis of previous flow control situations that had caused controller overload in Germany. Note that this value depends on the methods, procedures and equipment available in the ACCs, and hence applicable only to German ACCs. The experience of Switzerland with using the DFS model has also been positive because the Swiss airspace has equivalent methods, procedures and equipment. In addition, the DFS settings for the workload parameters have been validated extensively in Switzerland.

A major advantage of the DFS model is that the formula defining the workload is simple, clear and easy to understand by controllers, and incorporates explicitly more factors than simply the number of aircraft. However, some of its drawbacks include the considerable time spent on the selection of simulation scenarios, and the number of scenarios (typically four or five) to simulate, taking into account computational load and other resources. In Germany, it takes between nine months to a year from the identification of simulation scenarios to the start of operations in a sector. Real-time simulations are not often conducted in Germany, and only for major changes in German airspace. Given that Swiss airspace is significantly smaller than German airspace, the TAAM simulations take considerably less time. Therefore, although requiring significant resources to initially simulate scenarios with TAAM, so great is the efficacy of its results, that DFS rarely requires a subsequent RTS. This results in significant resource saving overall.

The relative assessment approach In both France and Canada, TAAM is used for fast-time simulations that are systematically conducted to assess and validate changes in sectorisation. However, in contrast to Germany and Switzerland, the emphasis is not to obtain a final capacity figure. Instead, the TAAM model is used to make relative assessments of workloads for different sectorisations. Typically, the current airspace is simulated to provide a baseline for subsequent studies, including the consideration of other relevant factors. A major reason why the model is not used to derive a capacity figure is its inability to reflect the workload experienced by controllers in certain situations. In the case of Canada, such a situation arises close to the US border, an area requiring a considerable amount of coordination between the two airspaces. In this case, the current parameters do not accurately capture the major coordination workload. As a result, extensive real-time simulations are required to assess workload and capacity. Both nations acknowledge the major cost implications involved.

In France, for example, real-time simulations are required for major airspace changes and the capacity assessment is then made in consultation with the controllers, based upon their experience of the airspace. For minor airspace modifications, e.g. flows and route changes, smaller scale RTSs are conducted in individual ACCs, primarily to train controllers. The extent of controller involvement in France in defining the capacity assessment process and the declared capacity from an early stage should be noted. Whilst controllers tend to be conservative with regard to capacity figures, the experience in France is that their involvement throughout the capacity estimation process is of interest to both parties, as it provides valuable feedback from the controllers to the ACC management. On average, both FTS and RTS take six months each in France, and, in Canada, both can take up to nine months.

The no workload measurement approach Japan and the US use the TAAM model to estimate airspace capacity, but do not carry out controller workload assessment. When planning new airspace sectorisations or routings, TAAM simulations are used to investigate whether a proposed airspace and route structure can handle a given level of air traffic. Airspace planners investigate whether a particular air traffic flow pattern can be handled by the sector, with given aircraft performance data, and based on geometrical separation parameters. The SIMMOD (Airport and Airspace SIMulation MODel) is also used to simulate airport and airspace operations without taking controller workload into account. Indeed, crucially, no controller workload assessment is made in the FTS modelling. The assessment of workload is only made during the RTS conducted after FTS. Together, the FTS and RTS process can take up to two years in Japan.

The FAA assesses airspace capacity using a variety of analytical and simulation modelling tools. However, controller task-load and workload is not formally assessed during these simulations. The FAA has undertaken such capacity assessment before using the TAAM model. In addition, as mentioned earlier, SDAT has also been used to assess workload in the past, but this is limited, since it cannot be used to estimate future workloads. A key point to note in this context is that for en-route capacity estimation, it is the maximum instantaneous count of air traffic in the sector that is important. This figure is not allowed to exceed 17 aircraft, and controllers usually verify the air traffic flow on the TAAM model to ensure this figure is not exceeded. Real-time simulations are conducted, in part, to validate the instantaneous count. The combined FTS and RTS process can take from 12 to 18 months.

3.3.2   *RAMS-based Methods*

RAMS is a discrete-event simulation model (EUROCONTROL, 1999a). It is used by several ATS providers as part of their capacity estimation process. The task base in the RAMS model includes 110 generic ATC activities, in five main categories: coordination tasks (WL_1); flight data management tasks (WL_2); planning conflict search tasks to determine ATC clearances (WL_3); routine RT communications (WL_4); and radar tasks (WL_5). Each generic activity is assigned to a user-defined working position, e.g. Planning Controller, by identifying the RAMS trigger event that indicates the execution of activity. Every ATC activity defined for RAMS requires a finite duration, to indicate the amount of time that the controller needs to carry out the task, and a time offset which identifies when that activity should be started. Throughout the simulation, RAMS monitors and records the internal events that occur and allocates actual ATC activities to 'simulated' staff, based upon these trigger events. The overall workload for a sector is simply the sum of the workloads for the tasks in the five categories, i.e.:

$$\Sigma W = WL_1 + WL_2 + WL_3 + WL_4 + WL_5 \qquad [2]$$

$$\Sigma W = \text{Total workload}$$

Of particular interest is the modelling of the conflict detection (WL_3) and conflict resolution (WL_5) tasks. Post-processing typically involves plotting the best regression line between sector entry and controller workload. The intersection of this curve with the overload threshold at 70 per cent workload (i.e. 42 minutes' effective measured work per hour) indicates the sector capacity. There are a number of different approaches to the use of RAMS, as illustrated in Figure 3.3.

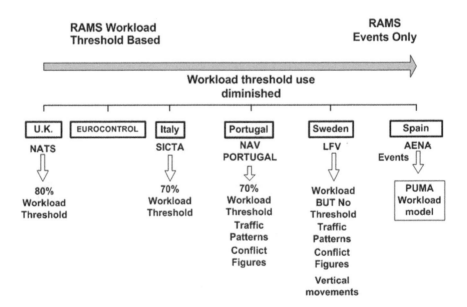

Figure 3.3   Basic approaches to capacity estimation using the RAMS model

Controller workload threshold approach The most common application of the RAMS model to airspace capacity assessment is based on the controller threshold approach. An example of this is in Italy, where the ANSP uses RAMS to estimate the capacity of new airspace sectors as part of the airspace design process. Capacity is determined in terms of the workload figures, and special emphasis is given to modelling and the analysis of conflicts. The EUROCONTROL methodology for assessing sector capacity is used by the Italian ANSP, i.e. capacity corresponds to 42 minutes in the hour (70 per cent) workload threshold. The 70 per cent value corresponds to 42 minutes' effective measured work per hour, leaving the controller 18 minutes for the general monitoring of traffic and for recuperation.

In the UK, NATS also uses RAMS for its airspace capacity estimations followed by post-processing using a custom workload post processor. There is a considerable degree of controller involvement in the process from task definitions and timings, to validation. However, NATS uses an 80 per cent workload threshold for determining capacity. The difference between the UK workload threshold and that of EUROCONTROL, is due to the introduction by NATS of a variable monitoring workload, depending on the number of aircraft in a sector. In both cases, RTSs are carried out after the RAMS simulation analysis to support the capacity estimates from FTS.

EUROCONTROL also uses the CAPAN methodology to assess capacity. In CAPAN, either the EUROCONTROL Airspace Model (EAM), a predecessor of the RAMS model, or the RAMS model itself is used to provide the FTS generator (engine). CAPAN then utilises an iterative process to saturate, or reduce, the ATC sector studied. It does this using a fully automatic process, and there is considerable involvement of ATC staff. Figure 3.4 shows the typical process by which CAPAN estimates the capacity for a sector given the data from a 24-hour simulation, either from RAMS or EAM.

Controller workload threshold, conflict and traffic pattern approach NAV Portugal uses the EUROCONTROL methodology for capacity estimation but only as part of a much wider process. This is because of the specific features of traffic patterns in Portuguese airspace, which renders the basic method inadequate. There are five major areas defined by their characteristic traffic flow: Canary Islands, Lisbon area, Faro area, Madrid area and Oceanic. Traffic patterns can vary considerably depending on the day of the week. Even during a particular day, hourly traffic distributions vary significantly. While RAMS is a powerful method, the experience of NAV Portugal is that of a complex and long process that relies on an estimation of controller workload and requires several simulation runs. Special emphasis is again placed on the RAMS simulation for conflict resolutions and workload differentiation in which close cooperation with controllers is required. Due to local variations in traffic flows, NAV Portugal also considers (in addition to workload) complexity issues (such as traffic patterns and conflict data) from the RAMS simulation to determine the capacity of en-route sectors.

Real-time simulations are used to validate FTS results. However, it is important to note that traffic samples in the RTS are limited to three or four exercises, due

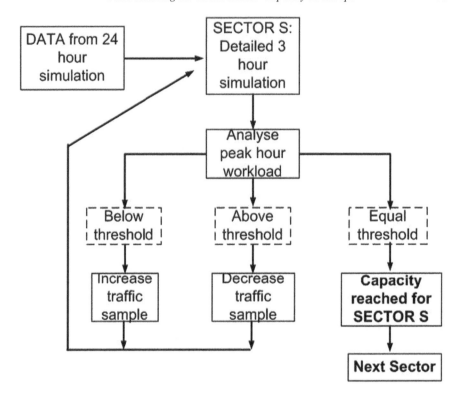

Figure 3.4   CAPAN approach to capacity estimation

Source: private communication, EUROCONTROL.

to the considerable time needed to prepare a simulation, and therefore do not represent all possible situations that may occur in a sector. Finally, and once the new airspace structure is open in the control room, the day-to-day operation permits a better adjustment of the first declared capacity value, taking into consideration a real traffic sample. It takes on average one year to implement a new airspace structure, with RTS taking four months.

Non-workload threshold, traffic and conflict patterns approach   This approach is employed by the Swedish ANSP, Luftfartsverket (LFV), and involves the use of relevant output data from the RAMS simulations to assess capacity, instead of the threshold-based methods. The data include the number of different tasks, calculated workloads, the number of climbs/descends, aircraft statistics and the number of flight-level changes. Operational controllers in the simulation area are then used to assess sector capacity. RTS, although ideal after RAMS simulations, is expensive, and this often precludes its use in Sweden. The tradition at LFV has been to greatly involve controllers in developing sector capacity assessment issues. Although RAMS can do much in the way of simulation, it cannot account for the 'human thinking' component and for this, LFV use experienced controllers from the airspace area to be simulated to provide procedures as input for the RAMS simulations. Controller activity data from real-time simulations are used to enhance en-route sector estimation.

Combined workload model approach Spain's ANSP, AENA, estimates airspace capacity using two FTS models in a unique combination. Initially, there is a RAMS simulation for the en-route regions. These provide events, such as flight entries and exits into a sector, the number of conflicts, and any other ATC activities. The events obtained from the RAMS simulations are then input into the PUMA model to determine workload, which is post-processed to determine its relationship with the number of movements. The capacity of a sector is then the number of aircraft such that the controller workload threshold is 70 per cent. The principle of the workload calculation is that different types of aircraft movement in the sectors (for example: for cruising aircraft and aircraft in vertical evolution, i.e. ascending or descending), require different tasks for their control – thus leading to different workloads. In addition, two tasks undertaken simultaneously (for example: resolution of a conflict whilst controlling a flight in evolution), require more work than if undertaken separately. The workload algorithm in PUMA provides penalties for this and allows extra workload to be incurred.

PUMA has working units as its output, the limit of which is based on frequent, repeated air traffic controller activities during a 60-minute period. AENA has conducted studies to analyse those activities that are more frequent in an average working hour in the ATC environment. These key events are simulated and weighted accordingly in the PUMA workload model. Different workload limits are used in PUMA for the planning, tactical and procedural[2] controllers' workloads, given their different tasks. AENA has been gathering data from Spanish ACCs since 1995 and thereby modifying and adapting the parameters in the PUMA workload algorithm. Real-time simulations are usually run after fast-time simulations, though this depends on the nature of the airspace changes. AENA expresses satisfaction with the methodology employed in the capacity estimation process, as it makes best use of the simulation models currently available.

3.4   Other Relevant Capacity Issues

During their analysis, Majumdar et al. (2005) encountered a number of other issues relevant to the en-route capacity estimation processes.

ATC Complexity: there was a general consensus from ANSPs interviewed on the need for an ATC complexity formula. Whilst Germany has the DFS model, which has proved its worth in German and Swiss airspaces, it appears that such a formula cannot easily be exported, as can be seen with the cases of France and Canada. Furthermore, the use of such a formula, avoiding the need for RTS, has the potential for significant cost savings. There are various projects

---

2   Procedural control is a method of providing air traffic control services without the use of radar.

underway in Europe (Hilburn (2004), for example) and the US, to derive this elusive complexity formula.

RTS issues (Canada): Whilst RTS simulations provide valuable controller experience in the capacity estimation process, they require considerable resources. Some ATC providers have attempted to improve the RTS aspect of capacity estimation. An example is NAV CANADA which has built its own real-time simulator, CAMSIM (Canadian Management Simulator). CAMSIM is familiar to all controllers in Canadian ACCs and has had the positive impact of cost saving on training. The proposed next step is to have CAMSIM remotely operated and to drive the simulation scenarios from the main simulation centre. This should remove the need to take controllers out of operational duty for a period of time.

Technology limitation on capacity estimation (Belgium): In Belgium, while controllers were able to cope with increased operational traffic planned after the CAPAN simulation exercises, the capacity of the ATS computer systems could not. In particular, the need for on-screen cleared flight level (CFL) inputs for level changes meant that, in the complex Belgian airspace, with multiple aircraft climbs and descents, the computers could not cope with the speed and number of inputs. Therefore, there is a need to also consider the capacity of the ATS system in future capacity estimations, a fact noted by other studies (Hudgell and Gingell, 2001).

Multi-model data transfer (the PITOT tool, Spain): AENA makes best use of simulation models currently available. However, this has the drawback in that there are multiple files and data conversions required from model to model. It is a time-consuming process to go from a typical day's air traffic, to the events from the RAMS simulation, and then to calculate workload in PUMA. The user thus spends a lot of time moving and changing files from model to model. Therefore, AENA is developing the PITOT (Process-based Integrated PlaTform for Optimal use of analysis Techniques) tool to automate the whole process. This will mean that file transfers between the various simulation models will all be automated.

Controller involvement: controller involvement is an essential part of reliable capacity estimation. However, the level of involvement can vary. At a basic minimum, controllers need to define their en-route tasks, timings and task frequencies in order to define the inputs into the FTS models and also to validate the simulation. There is a need for balance in controller involvement between obtaining the benefits of realism and validity, on the one hand, and resourcing costs and subjectivity biases, on the other.

3.5   ATC Complexity and Capacity

We have already noted that the workload experienced by controllers is affected by a complex interaction of a number of factors, primarily the situation in the airspace as determined by:

*   physical aspects of the sector, e.g. size or airway configuration;
*   factors relating to the movement of air traffic through the airspace, e.g. the number of climbing and descending flights;

- a combination of the above factors, which cover both sector and traffic issues, e.g. required procedures and functions.

The interaction between sector and air traffic features is a complex process, which can be termed as ATC 'complexity'. The nature of complexity is such that the variables involved interact to generate workload and these variables can be thought of as the primary (or source) factors affecting controller workload. The impact of these primary factors on controller workload can be mediated by secondary factors. It is reasonable to assume that as the ATC complexity increases, the workload of the controllers also increases. This rise in workload could have the consequence that controllers make errors in their task of enforcing separation requirements. Hence the analysis of ATC complexity factors can also be used in assessing safety. This is especially the case for sector design, to ensure that a particular sector is not designed in such a way that the workload for the controllers is too high. Figure 3.5 shows the relationship between ATC complexity, safety occurrences and workload.

EUROCONTROL's Performance Review Commission recognised the importance of traffic complexity when computing ATM performance. Complexity indicators are now computed on a systematic basis over a period of 365 days that can be used to benchmark ANSPs throughout Europe. These complexity indicators are based upon the concept of 'interactions', i.e. the simultaneous presence of two aircraft in a cell of airspace 20 by 20 NM and 3000 feet in height. The complexity score for each ANSP is determined by calculating the product of the amount of traffic in the airspace and a measure of the vertical movements and speed differences between aircraft (EUROCONTROL, 2006m).

There have indeed been numerous studies on ATC complexity factors: see Hilburn (2004) for an excellent review. However, it has become increasingly apparent that there is a significant need for the prediction of complexity-safety impacts when using complexity modelling tools, whether FTS, or other modelling approaches. Although taxonomies of complexity factors exist, these are often not broadly-based enough, and it is often unclear how they might interface with fast-time simulation tools which predict workload based on sector characteristics.[3] There is, therefore, a need to develop a consolidated list of complexity shaping factors (CSFs) that can impinge on workload and investigate whether these can indeed be utilised in a predictive fashion for new airspace considerations. If feasible, then research is needed to integrate such factors into tools that are being developed to predict workload, to ensure that future sector design remains safe.

3.5.1   Complexity Shaping Factors: a Taxonomy

A recent study by Majumdar and Ochieng (2007) used interviews of controllers to obtain, as subjective factors, a taxonomy of CSFs that affect workload. Face-to-face interviews were conducted, based upon a structured questionnaire, with

---

3   Based upon private communication with EUROCONTROL Experimental Centre.

SOURCE FACTORS   MEDIATING FACTORS   RESULT

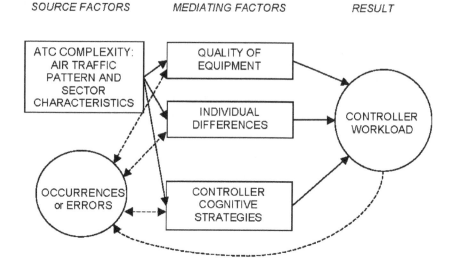

Figure 3.5   ATC complexity, controller errors and workload

Source: Modified from Mogford et al. (1993).

95 controllers – not just in Europe (44), but also in high traffic growth areas of Asia (48), as well as Johannesburg ACC (3), the busiest in Africa. Initially, these variables were determined using a literature review, and a series of detailed questions were framed to ascertain the impacts in areas such as: flight profiles, aircraft speeds, routes, intersection points and restricted areas.

The interviews were conducted in the appropriate Area Control Centres[4] (ACCs), i.e. en-route centres, in Europe and Asia. The choice of these ACCs was of importance, and those chosen encompassed:

- areas of high complexity and traffic density (e.g. Maastricht Upper Area Control Centre (Maastricht UAC) in Europe and Tokyo in Asia), as well as areas of less complexity (e.g. Oslo ACC);
- areas where complexity has increased rapidly in the past decade (e.g. Vienna ACC in Europe and ACCs in India).

Such a selection of ACCs allowed for the analysis of both high-complexity and rapidly-increasing complexity areas, as well as for a major geographical mix. Indeed, this mix of ACCs should ensure that the relevant complexity factors are obtained for their impact on workload, and represents one of the largest geographical mixes of controller interviews recorded. For each ACC, at least two controllers were interviewed to off-set the bias that a sole controller interview could have produced.

---

4   US readers, for example, may find it helpful to be reminded (from Chapter 1) that an ACC is the European analogue of an Air Route Traffic Control Center (ARTCC).

A taxonomy of over 50 complexity variables was created, as grouped by category in Table 3.5, which also shows whether such a complexity variable has previously been described in the literature on safety or airspace capacity.

3.5.2   Commentary on CSFs

Most of the variables in Table 3.5 are self-evident. Others, especially those relating to sector geometry, may not be readily apparent. For example, Figure 3.6(a) indicates a crossing point located close to the sector boundary with two entry points. This situation provides considerable difficulty for the controller in both coordinating entering aircraft and resolving conflicts between converging traffic. Figure 3.6(b) shows the angle of intersection of a route with the sector boundary in both a non-ideal and ideal mode. In the ideal case, the angle of intersection should be at right angles for ease of controller workload. Figure 3.6(c) outlines how sharp-edged sectors can pose a problem for controller workload with some controller difficulties in assessing the distances from the edges to the crossing point. Finally, Figure 3.6(d) indicates how a differential vertical split in two adjacent sectors of airspace can create extra coordination workload for the controller. Coordination with neighbouring countries could pose a major workload problem for the controllers, especially those with different procedures and technologies. Even with similar procedures and technologies, linguistic differences between controllers when speaking English, added to their complexity.

Weather poses a major problem for controllers globally and the topography of a country also plays a large role in this. In countries where sectors overlie mountain ranges, e.g. the Vienna sectors over the Alps or the Johannesburg sectors over the Drakensberg mountains, controllers find themselves more prone to changeable weather conditions and, in summer time, to the development of thunderstorms. In addition, it not solely the duration of weather conditions itself that affects controller workload. Controllers highlighted the fact that immediately following a period of bad weather, there tends to be severe workload as aircraft prevented from departing and landing during the bad weather request to do so (an issue intimately associated with slots – see Chapter 2).

CSFs unrelated to aircraft movements and sector geometry can also influence controller workload. Some of these factors can be dealt with by the ANSP of a country, e.g. the quality and range of the radar coverage and RT communications within the state. However, there are other factors within a country that are beyond their control, e.g. pilot compliance with controller instructions. Furthermore, there are factors that a country, or its ANSP, are much less able to affect. For example, Vienna ACC controllers noted that the air navigation charge regime of their neighbouring countries led to an increase in the number of aircraft overflying and undertaking complex movements in Austrian airspace.

In order to ascertain the importance of the complexity variables, the taxonomy list in Table 3.5 was then presented to airspace planners in the United Kingdom, Ireland, Denmark, Portugal, Austria and EUROCONTROL. The planners were asked to rate the impact of the complexity variables on their airspace based upon

Table 3.5   Taxonomy of complexity variables

Complexity variable	Existing use?
Traffic measures	
number of aircraft entering sector	yes
max number of aircraft instantaneously in sector	yes
frequency congestion measure	yes
clustering of aircraft in sector	yes
Traffic mix measures	
mix of descends and ascends	yes
mix of slow and fast moving aircraft	yes
mix of jet and turboprop aircraft	yes
mix of aircraft performance measures*	no
mix of aircraft equipage	no
Traffic speed mix – speed difference between:	
slow lead aircraft and trailing fast aircraft on same route	yes
slowest and fastest aircraft at same flight level entry point in specified period	yes
Entry and exit point measures	
number of entry points (weighted)	yes
ratio of entry/exit points	yes
geographical entry and exit points	no
clustering of entry and exit points	no
Routes measures	
number of routes in sector	yes
route miles flown	yes
number of intersecting routes in sector	no
number of routes which change direction	no
number of routes close to sector boundary	no
proportion of unidirectional to bi-directional routes	yes
number of intersecting bi-directional routes	no
route length	yes
Intersection points, reporting points	
number of intersection points in sector (weighted)	no
angle of crossing at intersection point	no
geographical location of intersection points in sector	no
clustering of intersection points in sector	no
number of compulsory reporting points prior to entry into sector**	no
Flight levels	
number of flight levels used in sector (weighted)	yes

Table 3.5 cont'd

Complexity variable	Existing use?
Neighbouring sectors	
number of neighbouring sectors (weighted)	yes
number of surrounding sectors with considerably more difficult procedures	no
number of neighbouring country sectors at capacity	no
differential vertical split of sectors with neighbouring country	no
number of surrounding Oceanic sectors	no
no. of surrounding sectors with different procedures required (weighted)	no
flight time through neighbouring sector[†]	no
number of sectors adjacent to neighbouring sectors requiring special work[‡]	no
Restricted/special area, military airspace	
location(s) of restricted area(s) within sector	no
volume of airspace restricted in sector	yes
military aircraft routes crossing civil aircraft routes	no
location of areas within sector in which slow moving aircraft	no
volume of areas within sector in which slow moving aircraft	no
Sector geometry	
sector volume	yes
sector shape (difference from regular polygon)	no
angles of routes intersection with sector boundaries	no
sharpness of sector boundary edges (angle) in relation to routes	no
angle of parallel flow to sector boundary	no
Weather	
topography of sector (can expose aircraft to bad weather)	no
location of bad weather regions in sector	no
clustering of traffic in sector following end of bad weather	no
Others	
pilot compliance with instructions	no
pilot experience/type	no
human-machine interface in control room	no
range and quality of radar	no
range and quality of RT communications	no
environment of control room	no
time of peak traffic period (e.g. circadian rhythms)	no
policy of neighbouring country	no

Notes to Table 3.5:

*   Especially in climb and descend.
** From Oceanic sectors (see also Chapter 1 for discussion on reporting points).
†   Indicates if primary sector has to undertake actions to prepare for adjacent sector (e.g. see next entry in table).
‡   Procedures in neighbouring sectors require special controller work in primary sector, e.g. RVSM (Reduced Vertical Separation Minimum) to CVSM (Conventional Vertical Separation Minimum).

their experience, using the following marking scheme: 3 for maximum impact; 2 for moderate impact and 1 for minimal impact.

It is apparent from Table 3.6 that the first eight factors (with the highest ratings) are those that relate to the aircraft, in particular, to vertical movement. The most important route features relate to the intersection of routes, whether unidirectional or bi-directional.

There is a need, therefore, to incorporate explicitly the impact of the CSF on workload in a predictive fashion for capacity and airspace design considerations. This will ensure that it is possible to estimate airspace capacity for future scenarios and that sector designs remain safe. This requires quantitative values for the CSFs, individually, in combination and at different levels.

It is possible to use FTS to help determine the functional links between the CSFs and simulated workload to enable the prediction of high-workload, unsafe situations. This is essentially the approach used by the DFS TAAM workload model, as per equation [1]. Another example of this can be seen in the use of panel data techniques to analyse the CSFs – such studies indicate the following variables are statistically significant (Majumdar et al., 2004):

- number of aircraft in:
  - continuous cruise
  - continuous climb
  - continuous descent
  - cruise-climb profile
  - cruise-descent profile
  - climb-descent profile
- total flight time
- aircraft speed mix
- number of flight levels
- number of exit points
- coordination measures.

As can be seen, most of these CSFs relate to air traffic variables, with fewer sector variables. One of the most important features of future research in the FTS approach to capacity estimation will be to further assess CSFs affecting workload and to develop a robust functional link between the CSF and capacity.

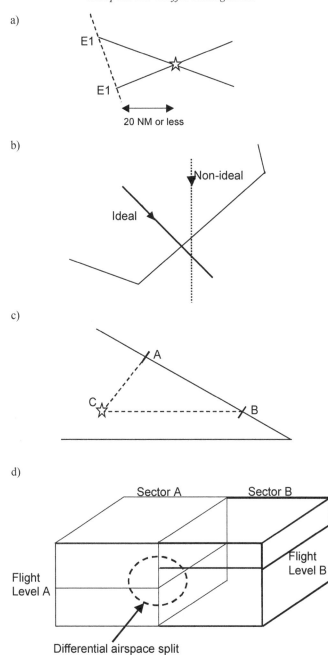

Figure 3.6   Four complexity variables relating to sector geometry

Table 3.6   ACC Taxonomy variables rated by impact

Variable	Average rating
Mix of descends and ascends	3.0
Pilot compliance with instructions	2.8
Range and quality of RT communications	2.8
Maximum number of aircraft instantaneously in sector	2.7
Number of aircraft entering sector	2.5
Frequency congestion measure	2.5
Clustering of aircraft in sector	2.5
Mix of slow and fast moving aircraft	2.5
Number of intersecting routes in sector	2.5
Number of intersecting bi-directional routes	2.5
Location of bad weather regions in sector	2.5
Time of peak traffic period (e.g. circadian rhythms)	2.5
Speed difference between slow lead and trailing fast aircraft	2.3
Number of flight levels used in sector (weighted)	2.3
Location(s) of restricted area(s) within sector	2.3
Clustering of traffic in sector following end of bad weather	2.3

3.6   The Future ATC System

The rapid rise in air traffic in both Europe and America, together with forecast growth, led the International Civil Aviation Organisation (ICAO) in the 1980s to recognise that the traditional ATC systems would not cope with such growth. ICAO's Special Committee on Future Air Navigation Systems (FANS) recommended the development of a satellite-based system to meet the future civil aviation requirements for communication, navigation and surveillance/ATM (CNS/ATM). In Europe, EATM[5] aims to implement policies of the ATM 2000+ Strategy, which involves the development of ATC/ATM strategies based on new technologies within the CNS functions of ATM (EUROCONTROL, 1991). Two key changes envisaged for en-route airspace are the introduction of real-time 4D navigation and the increasing delegation of responsibilities for control to flight crew, by the use of airborne separation assurance between aircraft, leading eventually to 'free flight' airspace.

EATM proposes a broad range of measures and technologies, progressively introduced, to keep pace with the anticipated changes. EUROCONTROL's Provisional Council has defined the following performance objectives:[6]

---

5   Formerly known as EATMP, and before that, as EATCHIP (see footnote 1).
6   Quoted from the website cited in footnote 1.

- to further improve ATM safety whilst accommodating air traffic growth;
- to increase the efficiency of the ATM network;
- to strengthen ATM's contribution to aviation security;
- to strengthen ATM's contribution to environmental objectives;
- to match capacity and air transport growth, towards the economic optimum in delays capacity.

In particular, as part of the future ATC/ATM service for Europe, aircraft will be controlled accurately and with high integrity in four dimensions with the aid of on-board and satellite navigation and communications technologies. Each aircraft will negotiate and re-negotiate a 4D flight plan in real time with the ground-based ATM system. This will provide airborne autonomous separation to give conflict-free tracks between origin and destination in the form of 4D profiles to be accurately adhered to by aircraft. (See also the discussion on navaids, RNAV and P-RNAV in Chapter 1).

Furthermore, moves towards free flight will result in pilots being relatively free to choose their preferred route between departure and destination airports. In Europe this will happen in stages, with increasing degrees of delegation of responsibility to flight crew. This move towards free flight, with pilots responsible for aircraft separation, will lead controllers and pilots to face several changes to their operations. For example, traffic will no longer be constrained to a fixed route structure with predictable congestion points (see Chapters 2 and 4) and controllers may need to be familiar with the whole airspace (as opposed to fixed flight profiles) to allow flexibility in the dynamic assignment of airspace to traffic. Such delegation will have profound influence on the roles and consequent tasks that pilots and controllers must carry out, not least with respect to monitoring. This, in turn, will impact on their workload and hence capacity in en-route European airspace[7] (see Table 3.7). Currently, controllers know the essential intent of aircraft through the concept of clearances based upon the flight plan. In theory, greater aircraft autonomy provided by means of advanced airborne technologies, can only improve safety and efficiency of today's system. For example, transferring aircraft separation workload from a small number of controllers to a large number of pilots should reduce controller workload. However, aircraft-based conflict detection and resolution must be reliable. Therefore, there is a need to understand the implications of such developments for en-route airspace capacity.

The future ATC environment will require major modifications in estimation methodologies with the need, for example, to consider capacity as determined by a combination of controller and pilot workload. Furthermore, the change in nature of the controllers' tasks to more of a monitoring role also poses problems for current methodologies as, currently, monitoring is the least well modelled of the controllers' tasks in FTS. This will require major changes to the FTS tools.

Therefore, when considering the future ATC environment, more emphasis is likely to be placed upon RTS to obtain the controller data, e.g. tasks, frequency

---

7   The potential environmental impacts of such reorganisation are discussed in Chapter 5.

and timings. As a consequence, until FTS models develop to adequately capture the new control environments, there will be a need for expensive real-time simulations.

Turning finally to the issue of new tools, it is apparent that the introduction of increased automation for controllers in terms of decision support tools (DSTs) will lead to increasingly changed methods of working. An example of this is the First ATC Support Tools Implementation (FASTI) Programme (EUROCONTROL, 2006c). The scope of FASTI encompasses the implementation of Medium Term Conflict Detection (MTCD) and enablers such as Monitoring Aids (MONA) and System Supported Coordination (SYSCO). Trajectory Prediction (TP) and the Human Machine Interface (HMI) are critical to the performance and use of FASTI tools. Therefore, one of the major aims of FASTI is that, with the assistance of DSTs, the Planning Controller (PC) will attempt to reduce the workload of the Tactical Controller (TC) – the latter being a determinant of capacity in Europe, as already mentioned. MTCD and MONA will assist the PC in identifying problems earlier than at present, with, hopefully, major improvements in airspace capacity. Earlier identification of problems would permit the PC to intervene, assess and resolve *some* problems, thus balancing the workload of the TC. Crucially, ATC planning and management will be conducted strategically using FASTI tools to generate nominally conflict-free trajectories. An initial attempt has recently been made by EUROCONTROL to model the impact of the FASTI toolset on airspace capacity by means of FTS, using the Maastricht upper en-route airspace environment,[8] with promising capacity effects predicted.

In future, further refinements to such analyses should consider: new methods of controller task sharing; enabling the PC to check the TC's predicted workload in a given short-term time frame, e.g. ten minutes; better modelling of increased monitoring tasks (which will be a result of the increasing use of automation, but are currently less well modelled by FTS); and consideration of trust issues. There is a clear future need for capacity workload thresholds to change, in order to encompass these factors.

3.7   Conclusions

As air traffic grows it is important to be able to assess accurately and reliably en-route airspace capacity, especially in high-density traffic areas such as those found in Europe, where the limiting factor on capacity is controller workload. The advent of new technologies and procedures means that the importance of accurate and reliable capacity estimates is increased.

This chapter has shown that fast-time simulation of controller workload is most commonly used to assess sector capacity. Whilst there are two major controller workload models, TAAM and RAMS, they are both based upon similar principles of discrete-event simulation modelling. Both these models have

---

8   Private communication from EUROCONTROL.

Table 3.7   Possible changes in duties due to increased delegation

Delegation	Air traffic controllers	Pilots
Limited	In charge of both identification of problems and solutions	Only implement solutions and monitor
Extended	Identification of problems	Identification and implementation of solutions and their monitoring
Full	Monitoring	Responsible for all tasks related to separation: identification of problems and solutions, implementation and monitoring

Source: Based on The CAST Consortium findings (1998).

developed formulae to assess sector capacity based upon variables extracted from the simulation output.

There is a varied level of post-processing carried out after a FTS. Some ANSPs use a formula-based approach, e.g. in Germany. A number of countries choose not to use a formula alone to assess sector capacity. Rather, they use the workload estimation together with other traffic and conflict factors, to enable a panel to decide on sector capacity. Perhaps there is a case to move towards a formula to reliably assess capacity and thereby avoid the time-consuming process of panel discussions. However, as the use of the TAAM-based DFS workload formula showed, transferability is a major concern. This is an indication of the difficulty of deriving a generic complexity formula. Even the differences in traffic patterns and sector geometry between ACCs in one nation can be considerable.

Controller involvement throughout the process is essential and all nations surveyed accept this. However, its extent needs to be managed to prevent the process from being too lengthy and expensive. Major controller involvement and resources are needed for RTS following FTS. It does appear that nations using a validated formula, e.g. Germany and Switzerland, do not need major RTS. Those countries not using even a threshold value of controller workload for the FTS require major RTS. Therefore, the effort required in developing and using a formula may lower costs in the long run.

An increasing requirement for capacity estimation by FTS is the development of a robust complexity formula for post-processing. Whilst there has been considerable research on complexity factors, there is still a need to develop a robust formula linking CSFs to workload and, consequently, capacity. An additional factor to note is that most such factors currently analysed tend to be movement-related rather than sector-related. In addition, rarely are non-linear combinations of factors analysed in these formulae.

For the future autonomous ATC/ATM environment, major revisions will be required for the FTS methodology given the major changes in roles and tasks for controllers and pilots. Primarily, there will need to be an assessment of the tasks associated with each discrete event of the FTS, followed by an understanding of how pilot and controller tasks interact. Even in a non-autonomous environment with increasing use of DSTs, new methods of working need to analysed and modelled. These include such factors as greater cooperation between the PC and TC, greater monitoring tasks for controllers and the introduction of treatments of trust and judgement to the output of DSTs. Given the current difficulties of using controller complexity formulae in the ATC environment, deriving such formulae for the future environment is one of the great challenges for ATM.

Chapter 4

The Management and Costs of Delay

Andrew Cook
University of Westminster

4.1   Defining the Concepts of Delay

Our focus in this chapter will be on the cost of delays to airlines. Although delays cause costs to other bodies, such as airport authorities, air traffic control (ATC), handling agents, and, not least, to the passengers themselves, limitations of space require a concentration here on the airline perspective. As we will explore in some detail, passenger delay generates a high proportion of these costs, yet this is often the least well quantified aspect of the problem.

Various definitions of delay and punctuality are used in the air transport context. For the purpose of this chapter, in particular where costs are allocated to delays, our definition will be the off-block/on-block time of an aircraft relative to the airline's published schedule. This is the most transparent metric, and the one with the most direct relevance to the ultimate customer of the system: the passenger.

The industry frequently has a rather stronger focus on departure delay rather than arrival delay, which is somewhat at odds with both the passenger perspective and airline marketing strategies: often promoting service levels in terms of arrival punctuality, although some carriers are now selling pretty much on cost alone.

Nevertheless, delays are bad news for airlines – the cost of delay hits airlines twice: both in the contingency planning of a schedule (the 'strategic' cost of delay), and then again, when dealing with actual delays on the day of operations (the 'tactical' cost of delay).

Although it is universally accepted that both these types of costs are real, and often very large, they are generally only poorly understood quantitatively, especially the strategic type. This chapter sets out to explore how these costs arise and to quantify them.

We will refer quite often to the process of Air Traffic Flow Management and 'ATFM slots'. These slots may be described as 'departure permission slots' issued by air traffic flow managers to optimise the flows of flights in conditions where demand is in excess of capacity. These slots, and the processes through which they are managed, were described in detail in Chapter 2. They are distinct from 'airport slots', which we will discuss in the next section.

4.1.1   Punctuality – a Question of Perspective

When reporting on punctuality, the air transport industry usually neglects shorter delays, for example counting departures and arrivals no later than 15 minutes with respect to the schedule, as 'on time'. The percentage of flights arriving no later than 15 minutes after the scheduled time is a common key performance indicator, both in Europe and in the US. In some ways this is reasonable, in others it is unhelpful.

Let us consider why it may be considered reasonable. Firstly, evidence suggests that, in a general transport context, passenger awareness of scheduled arrival and departure times is relatively imprecise, with a tendency to neglect small delays (Bates et al., 2001). The airline passenger may well agree that 0730 compared with 0715 is good enough to count as 'on time'. Secondly, air transport works within margins of tolerance with respect to timings. At airports, it is not unusual to have between five and ten flights scheduled to depart or arrive at exactly the same time, whereas in actual fact this is not possible, due to airport capacity limitations (particularly runway constraints). Exactly how these 'simultaneous' departures are managed on the day, is a matter for the operational practice of the airport and ATC.

Furthermore, as already commented upon in Chapter 2, the ATFM departure window is 15 minutes wide[1] if ATFM regulations are in place (i.e. if the flight has been given a departure slot, or CTOT), and 30 minutes wide otherwise (when the flight is said to be 'unregulated'). Whereas many facets of ATM work according to extremely precise control mechanisms, for practical purposes, there are necessary tolerances associated with certain timings. These allow for the unpredictability of taxi out times or winds, for example.

On the other hand, it may be argued that counting delays of up to 15 minutes as a 'punctual' arrival is less helpful in two major respects. Although 15 minutes may often be a relatively short delay from the viewpoint of the passenger, it could be far more important if this causes a missed connection. As airlines try to squeeze ever greater efficiency out of their networks, the Minimum Connect Time (MCT; see Chapter 6) is often very short, and the loss of 15 minutes could well be enough to cause an onward connection to be missed.

How delays are documented and analysed varies from airline to airline, from reporting body to reporting body, but counting delay from the first minute can only help the general cause of transparency, especially as these have a tendency to accumulate. Neglecting or rounding-off 'low' values makes causal diagnosis more difficult. ATFM delays, however, *are* counted from the first minute of delay (see later).

To further complete the picture of the framework within which delays may be defined, it is necessary to first return to the issue of 'airport slots'. Twice each year, since 1947, in advance of the summer and winter schedules, the IATA Schedules Conference takes place to allocate 'airport slots' to the airlines. To

---

1   From –5 to +10 min, although around one in five take-offs actually occurs outside this window.

manage the imbalances between airport demand and capacity in Europe, the EU has established a set of common rules[2] for the allocation of such slots. It defines an airport slot as:

> the permission given by a coordinator in accordance with this Regulation to use the full range of airport infrastructure necessary to operate an air service at a coordinated airport on a specific date and time for the purpose of landing or take-off as allocated by a coordinator in accordance with this Regulation.

According to the level of congestion, three levels of coordination for airports are defined, as shown in Table 4.1. Slots are allocated from a common 'pool' of unallocated slots, according to transparent rules, and referring to local guidelines. If a requested slot cannot be accommodated, the nearest available slot is suggested. Slots may be freely exchanged (if local and/or national regulations permit this), one for one, between airlines. This is in contrast to ATFM slots, which are very rarely exchanged (although some trials are looking at this).

Table 4.1   Airport slot coordination

EU term	IATA term	Usage
Coordinated*	Level 3	A slot must be allocated to the airline by a coordinator. 'Coordinated' is a legal status, set (or removed) by the state. Used for most congested airports.
Schedules-facilitated	Level 2	Used when there is potential for congestion during certain periods; voluntary cooperation with a schedules facilitator
Non-coordinated	Level 1	Low congestion. Airline deals directly with the airport/handling agent

*   In Europe, the often encountered term '*fully* coordinated' is strictly speaking a misnomer. An airport is either 'coordinated' or it is not. In rare situations, an airport could be designated as coordinated for a short period, whilst capacity issues are resolved.

At the airport, special rules exist to protect 'grandfather rights' (historical operation at a particular time) and to ensure new entrants are not excluded from an airport due to existing operators – 50 per cent of new slots must be given to new entrants, if such demand exists. Both schedules facilitators and coordinators are obliged to monitor the conformity of airlines with the slots allocated to them. Failure to use a slot for at least 80 per cent of the scheduled period, may

---

2   Council Regulation (EEC) No 95/93 (18 January 1993).

lead to the slot being lost (according to Council Regulation (EEC) No 95/93). Furthermore, the local schedule coordinator may also impose fines on a carrier taking off or landing outside the allocated slot time, if this is deemed to constitute 'slot abuse'. The airline may also lose the slot on this basis. Obviously, there is a degree of tolerance observed with regard to delays: flights very often have to arrive or depart outside the agreed airport slot, due to delays on the day of operations. An airport authority would very rarely refuse a flight due to a delay, but there is the additional incentive to operate according to these slots, so as not to lose them, as they are often very highly valued commodities, with reports that airlines have even intentionally operated non-profit making flights in order to maintain the right to the airport slot.

The airport slot time is usually the same as the time published in the airline's schedule, in third party timetables and databases (such as those of the Official Airline Guide) and in Computer Reservation Systems/Global Distribution Systems (CRSs/GDSs – such as Galileo and Sabre). Although minor discrepancies may occur here and there, these sources are usually harmonised. Airlines may, however, file a slightly different time in their flight plans, to better suit their operational requirements, whilst keeping a more marketable time in their advertised schedules (e.g. on the hour departure times on a given route from the home base at 0700, 1000, 1300, 1600 and 1900).

All in all, these varying definitions of timings lead to certain ambiguities, for the unwary, regarding the definition of delay. These are unlikely to cause any severe problem in understanding delay, but it is necessary to be aware of the differences, and potential shortcomings of using one type of metric, as opposed to another, in order to steer a clear path through the mechanisms of allocating costs to delays.

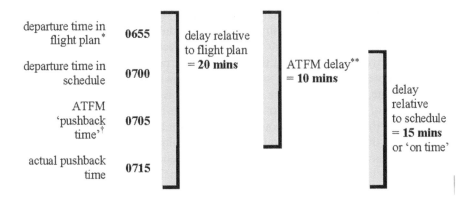

*   Estimated Off-Block Time (EOBT), see Chapter 2.
** To be precise, measured relative to the *last* take-off time requested by the aircraft operator.
†   Not actually specified as such, but implied from CTOT (taxi time + line-up time etc), see Chapter 2.

Figure 4.1   Delay as a relative concept

Figure 4.1 thus demonstrates how different definitions of delay might be used. The definition used in this chapter, for the purposes of calculating the costs of delay, is that shown on the right-hand side of Figure 4.1. The need to operate to the published schedule drives airlines' operations, their desire to recover delay and the basis on which passengers cost the airlines money when they are delayed.

Although, as we shall see later, ATM causes a relatively small proportion of the delays in Europe, much of the focus of delay management is on the ATM process, largely because this is an interventionist mechanism, such that it is open to redesign and adaptation, unlike exogenous factors such as the weather. Whether this ATM focus is misplaced, is a moot point.

This section has highlighted a number of necessary tolerances associated with the management of flights. These tolerances have concomitant costs, incurred not only by airlines, but also by the airport and ATC.[3] They also contribute to a reduction in predictability, which is examined next.

4.1.2   Tactical and Strategic Costs of Delay – a Question of Predictability

Airline delay costs arise from a lack of predictability. Flight delays are caused by a number of reasons, from passengers arriving late at the gate, aircraft 'going technical', the unavailability of a free gate at the airport, through to ATC and ATM issues. No matter what the cause of the delay, the result of an aircraft pushing back 30 minutes late will usually have the same downstream cost consequences, whether it was caused by a late passenger or poor weather.

However, much of the cost is locked into the schedule in advance: this does not appear in the airline's accounts, and is largely unknown, but still costs the airline months before the flight. This is the strategic cost of delay. Airline schedulers know that delays are bound to happen. They know this in advance, but they do not know the magnitude of the delays, nor when they are going to happen. If the scheduler knew in advance that flight AB123, scheduled to leave on its first leg of the day from its home base at 0730, was going to leave 15 minutes late and be subject to arrival management at the destination airport of another 15 minutes (such as the well known arrival stacks at Heathrow), thus arriving a total of 30 minutes late, his task would be a lot easier. In this (unrealistic) scenario, the tactical costs of delay would no longer exist, whereas many of the strategic costs of managing these 30 minutes would still impact on the airline (such as under-utilisation of the aircraft and planning for the holding fuel).

Strategic and tactical delay costs are interdependent (as will be explored quantitatively later). The more an airline tries to anticipate delay in advance, the more likely it is able to deal with it (more cost effectively) on the actual day. Thus, strategic delay planning is used to off-set the impacts, and costs, of tactical delays. For example, airlines will usually (but not always) incorporate 'buffers' ('padding') into their schedules, in order to better cope with delays. Other contingencies include extra staffing and spare aircraft. These costs are incurred at the planning

---

3   For example, by causing bunching or peaking of flows, which may require additional controller/staff numbers to manage the traffic.

stage, and are generally not readily escapable. Buffers cost money since they often lead to sub-optimal schedules. EUROCONTROL (2006a) has estimated that approximately €1bn per annum could be saved if half of the European schedules could be compressed by five minutes. The strategic cost of delay to the airline is perhaps best described as an opportunity cost.

For both airline planning purposes, and performance assessment, more information is needed than just average delays. Knowing that network average delays are ten minutes in duration is not as helpful as knowing that the delay on a particular aircraft's first leg of the day is three times as likely to be greater than 30 minutes, as that on the final leg. In scheduling operations, airlines will often build in larger buffers (to cope with longer delays) on earlier legs, as these typically have greater operational impact, since there is clearly more scope for these to propagate throughout the day, resulting in delays on subsequent legs, and thus compounding the original problem. This primary delay might also affect other aircraft, both of that operator and alliance partners, which may be held at-gate for connecting passengers. These 'knock-on' delays are referred to as 'reactionary' delays and may be defined as all delays which may be directly attributed to an initial, causal or primary delay. They might be experienced by the causal aircraft itself (rotational reactionary delay), or by others (non-rotational reactionary delay). Table 4.2 further illustrates the types of delay cost discussed thus far.

Table 4.2   Definition of four basic types of delay cost

Delay cost type	Definition	Example	Example cost
Strategic level	Resources committed at planning stage as advance contingency for delays	Airline putting buffers into schedules	Under-utilisation of aircraft
Tactical level	Costs incurred due to actual delays on the day of operations	ATFM slot delay due to airport capacity constraints	Re-booking delayed passengers
Rotational reactionary	'Knock-on' delay costs to the same aircraft that causes an original delay	Aircraft X: 60 mins late inbound, so same delay next leg	Passenger on aircraft X compensated
Non-rotational reactionary	'Knock-on' delay costs to other aircraft in the network	Aircraft Y waits 60 mins for aircraft X	Passenger on aircraft Y compensated

A full review of ATS performance metrics by Boeing (1999) included categories for 'delay' *and* 'predictability'. An obvious metric for predictability, is delay variance. Consider an example discussed in Chapter 6, where a hub airport operates 'waves' of arrivals and departures – here it would be better if all aircraft were to arrive and depart with the same (smaller) delay, rather than just having a few arrivals with a very long delay. Both scenarios could produce similar average delays, whereas the latter situation would be reflected in a higher variance.

Wu and Caves (2002), in an optimisation of schedule reliability for aircraft operations based on schedule and punctuality data from an (undisclosed) European airline, comment that whilst the mean delay can be easily produced in analyses, this index is of 'little help when an airline is attempting to investigate the potential bottlenecks in aircraft rotations because it only reflects a part of operational characteristics'. They go on to comment that the (statistically) 'expected' delay of an aircraft rotation has the advantage of considering stochastic effects of delays and the probability of occurrence.

Furthermore, Hansen et al. (2001) have modelled, for ten US domestic airlines, various performance metrics with cost impacts. Their conclusions 'challenge the prevailing assumption that delay prediction is the most important benefit' of AT(F)M enhancements, with 'irregularity' and 'disruption' factors having the strongest cost impacts. The seven performance metrics modelled were highly intercorrelated, and for this reason, principal component analysis was used to identify a set of factors which were linear combinations of the original variables (such as delay magnitude). The authors concluded that if forced to choose a single metric to track the cost-driving dimensions 'irregularity' and 'disruption', it would be flight cancellation, rather than average delay per flight. In terms of investment in AT(F)M enhancement, therefore, they conclude that measures preventing the serious disruptions which lead to cancellations would be better than those leading to incremental delay reductions.

In another US model formulated using a large hub-and-spoke data set, Bratu and Barnhart (2004) propose 'passenger-centric' metrics (challenging the flight-based, 15-minute performance indicator currently in use), stating that the average passenger delay[4] was 1.7 times greater than the average flight-leg delay. The passenger-centred delay values included the impact of cancellations, which had a considerable effect on them.

The air transport industry in general, and ATM in particular, has access to delay, booking, origin-destination and movement data on a sheer scale which other transport sectors can only dream of. Regarding data on delay, there is now greater attention paid to complementary metrics, in addition to reporting average delays only. For European ATM, one of the most useful general (data) sources is EUROCONTROL's annual *Performance Review Report* (PRR), published by the Performance Review Commission.[5] Since PRR 2005, this document cites variances as a measure of predictability and explains that 'predictability is also one of ICAO's Key Performance Areas. It addresses the strategic ability of airlines and airports to build and operate reliable and efficient schedules, which impacts both their punctuality and financial performance' (EUROCONTROL, 2007j).

Whilst average delay statistics are indeed better when complemented by other metrics, it has also been observed that these metrics are intercorrelated,

---

4   For their August 2000 data.

5   PRRs are published around April–May each year, after consultation with stakeholders, and report on the previous year.

i.e. reducing the average delay is likely to be associated with a reduction in the variance, too. EUROCONTROL has taken the step of introducing a measurable target for average en-route ATFM delays. Although this was not met in 2006, the Performance Review Commission (ibid.) states that:

> As the present en-route delay target is commonly accepted, proven to be achievable, and widely considered as a satisfactory performance level, the PRC proposes that this system-level target (1 minute/flight) be maintained for the coming years.

Airlines, for their part, vary in sophistication regarding the extent to which various data are used at the strategic, pre-tactical and tactical levels (as defined in Chapter 2). Historical data may be used to build schedules based around past performance, for example, knowing that Flight AB123 at 0730 arrived 30 minutes late, 25 per cent of the time last winter season, may well induce the scheduler to adapt the schedule for the next winter season. Pre-tactically, good flight planning systems allow the user to apply statistical conditions to explore different routing options, for example by imposing likely delays which may be caused by weather and/or by arrival management at the destination airport. These plans may then be updated tactically using live meteorological data and CFMU data (see Chapter 2). This allows better fuel planning, and better tactical management of the schedule on the day of operations. The allocation of buffers and turnaround times at the schedule planning stage has an important impact on the tactical stage.

4.1.3   Buffers and Turnarounds

The concept of schedule contingencies, or 'buffers', has already been introduced. This definition will now be refined. Consider a simplified example, of a service operating between airports A and B. Assume that the off-block time from A to B is 105 mins, whilst the corresponding time for the return leg (B to A) is 120 mins – routes may often have asymmetries due to permitted routings and/or prevalent winds. Further assume that the turnaround time for the aircraft is 60 minutes at B, and 65 minutes at A. This is the time it takes to deplane the inbound passengers and their bags, service and refuel the aircraft, then board the outbound passengers and load their bags. We are defining here the turnaround time as the *actual time required* to carry out these processes. An example timetable is shown in Table 4.3.

Table 4.3 shows two types of buffer on leg 1. 'Off-block' buffer is added to the time allocated to get from gate A to gate B. It is designed to absorb any off-block delays, such as during taxi, line-up and runway sequencing, and all airborne delay (such as arrival management). Off-block buffer might be specified by the airline in terms of ground-based and airborne components, to determine how much fuel the aircraft uplifts at A.

In theory, having arrived at B, the aircraft could return on leg 2, back to A, at 1030, since it arrived at B at 0930 and the turnaround time is 60 minutes. However,

Table 4.3   Timetable illustrating buffers

Leg	Scheduled departure	Off-block buffer	Scheduled arrival	Turn-around time	Slack time	At-gate buffer
1	dep. A: 0730	15	arr. B: 0930	60 mins	0 mins	15 mins
2	dep. B: 1045	15	arr. A: 1300	65 mins	0 mins	10 mins
3	dep. A: 1415	15	arr. B: 1615	60 mins	10 mins	5 mins
4	dep. B: 1730	15	arr. A: 1945	65 mins	10 mins	0 mins
5	dep. A: 2100	15	arr. B: 2300	(out-stationed overnight)		

in this example, the airline has added an at-gate buffer of 15 minutes to allow for delays incurred at B and a punctual departure at 1045, on leg 2.

Turnaround time depends largely on the size of the aircraft. It is driven by factors such as refuelling times, scheduled maintenance checks, whether freight is carried and the method of passenger handling: how many doors are used, whether at-gate or remote, if airbridges are employed and whether boarding passengers have pre-allocated seats. Turnaround times for low-cost and traditional airlines[6] are generally getting closer, as traditional carriers seek increasing aircraft utilisation.

At-gate buffer is defined as additional time built into the schedule specifically to absorb delay whilst the aircraft is on the ground and to allow recovery between the rotations of aircraft. However, other factors may prevent the schedule from being perfectly efficient. After an inbound flight is scheduled to land, it may be necessary for it to wait before its next leg in order to allow for connecting passengers to reach the aircraft, or for a crew change. The availability of airport slots may also be an issue. We may identify this as 'slack time': these waits are imposed upon the airline by factors essentially exogenous to the scheduling of the A-B/B-A legs.

In the example shown, no at-gate buffer is employed between legs 4 and 5, as the airline prefers to reduce the risk of missing a night curfew in place at B. The actual amount of buffer incorporated into a schedule will depend partly on the absolute importance of punctual operation to the carrier. It is a question of balance. For example, low-cost operators may be more inclined towards having low buffers, or even none at all, wishing to extract the maximum possible utilisation out of the aircraft, operating in a highly cost-driven market.

A traditional airline is more likely to market itself in terms of punctual performance, in addition to depending on connecting flights (e.g. within an alliance network). It is particularly costly to such airlines if feeder flights into hubs are

---

6   See Chapter 6, footnotes 1 and 2, for working definitions of 'low-cost' and 'traditional' airlines

delayed, as these may cause the maximum amount of disruption in terms of having to hold the connecting flights, re-book passengers who miss connections, and/or compensate passengers who are consequently delayed. These costs are compounded by the fact that they are often earlier in the morning, may carry relatively high proportions of higher-yield passengers and also cause greater reactionary delays. For these airlines, the cost of unpunctuality can be particularly severe and outweighs the opportunity cost of reduced utilisation – they tend to opt for larger buffers, especially earlier in the day.

Some airlines express concern that higher off-block buffers mean that the longer block-times in the CRS/GDS push the flight further down the listing, thus reducing sales. However, since agents, public and corporate, are more likely to filter flights by either cost, departure time, or arrival time, it seems unlikely that this effect alone is very great. As more and more volume is sold through websites, either directly or through third parties such as Expedia and Opodo, this effect is likely to diminish further.

In summary, the science (or art) of using buffers varies from airline to airline. Some do not use them, others apply statistical modelling methods to optimise the entire network. Where an airline systematically relies on arriving earlier than scheduled, or reducing the turnaround time, to help manage delay, these practices are effectively using buffers, whether they are formally called this or not. Table 4.4 summarises some of the key advantages and disadvantages of the use of buffers.

Table 4.4   Main advantages and disadvantages of buffers

Advantages	Disadvantages
Reduces costs incurred due to delayed passengers	Decreased aircraft utilisation
Reduces additional crew costs, e.g. overtime payments	May incur additional charges for occupying/waiting for gates
May increase market share as a result of improved punctuality	May reduce market share due to drop in CRS position
Improves arrival punctuality	May involve over-hedging fuel
Improves network stability	May increase the risk of compromising a night curfew
Helps prevent aircraft from missing ATFM slot on next leg	

4.1.4   An Overview of European Delay Statistics

Having defined a number of operational parameters, and before we move on to an estimation of the costs of delay, we are now in a position to take stock of some key delay statistics for Europe, in 2006. In that year, on-time performance of scheduled air services deteriorated for the third consecutive year, with 'late' arrivals (i.e. arrivals more than 15 minutes later than scheduled) increasing by a quarter since 2003, reaching 21.4 per cent. Late departures (using the same definition relative to schedule) also increased by one third in the same period, these being largely attributable to increases in turnaround delays and reactionary delays: these together forming around nine out of ten of such delays (EUROCONTROL, 2007j). With arrival delays clearly being correlated with departure delays (although more research[7] is needed into the specifics of this, in particular: deterioration and amelioration relative to schedule, after pushback), it is clear that most of the arrival delay in Europe is attributable to non-ATM causes.

However, a strictly accurate allocation between ATM and non-ATM related causes is not as easy as it may seem. Not all ATM delays are attributable to 'pure' ATM causes (such as flow management based on sector capacities), but these restrictions may be caused by weather (causing about half of the ATFM delays at airports, in fact, often affecting several airports at the same time) and other events which ATM does not control. Potential complications associated with the definitions of turnaround delays and reactionary delays will be discussed later.

Nevertheless, the figures collated and published by EUROCONTROL are a valuable resource for the industry, particularly in terms of monitoring trends. As we saw in Chapter 2, the ATFM delay of any given flight is attributed to the most constraining ATC unit affecting the filed flight plan. This could either be an en-route unit (generating an en-route regulation) or the departure or arrival airport (generating an airport regulation).

As will be discussed in Chapter 8, 1999 was a challenging year for European AT(F)M, due in large part to the Kosovo Crisis and higher than forecast demand. However, en-route ATFM delays improved from then on, through to 2004, then increased in 2005 and again in 2006. The increase was somewhat lower in 2006, despite this being a year with quite strong traffic growth (4.1 per cent), special events (such as the 2006 World Cup, in Germany) and high weather-related delays. Growth in 'low-cost' aviation in the same period was particularly marked, at 24 per cent (*ibid.*).

In 2005, there was an increase in total ATFM delays of 17.6 per cent on the previous year, followed by a further increase of 4.6 per cent in 2006. The increase in 2006 was due to en-route ATFM delays, as airport ATFM delays decreased slightly in this year.

---

7   A French airports' case study (EUROCONTROL, 2003b) concluded that arrival delay was 80 per cent dependent on departure delay and 16 per cent on load factor.

Table 4.5   Some key delay statistics for 2006

Annual traffic growth		4.1%
Flights with arrival delay > 15 minutes		21.4%
Flights with departure delay > 15 minutes		22.7%
... of which, due to:	... en-route ATFM regulations	1.6%
	... airport ATFM regulations	1.1%
	... turnaround delays	10.2%
	... reactionary delays	<u>9.8%</u>
		22.7%
Primary departure delays (> 15mins)[a]	... en-route ATFM regulations	12.4%
	... airport ATFM regulations	8.5%
	... turnaround delays	<u>79.1%</u>
		100.0%
Average en-route ATFM delays	actual (per flight)	1.4 min
	target (per flight)	1.0 min
Increasing en-route ATFM delays mainly caused by	'specific circumstances'[b] weather staffing issues	
Most constraining ACCs	Zagreb, Zurich, Madrid, Warsaw, Barcelona and Prague (accounting for 65% of all ATFM en-route delays)	
Airports where ATFM delays improved	Istanbul (IST), Zurich (ZRH), Amsterdam (AMS) and Munich (MUC)	
Airports where ATFM[c] delays were noted to 'remain a concern'	Vienna (VIE), Milan-Malpensa (MXP), Rome (FCO)	

a   Obtained by removing reactionary delays from previous list and re-scaling.
b   Such as the postponement of Swiss UAC implementation.
c   I.e. aerodrome /ATC capacity-related ATFM delay.

Source: adapted from EUROCONTROL (2007j).

4.1.5 Strategic Management of Delay at the Airport

Table 4.5 shows that, in 2006, primary delays of over 15 minutes originated principally (79.1 per cent) from turnaround processes. These data are derived from delay categorisations, using standard IATA delay codes usually recorded at dispatch. Turnaround delays are intended to capture the corresponding operations at the gate, which have been described above. However, it is inevitable that some level of ambiguity arises. An aircraft arriving late at gate X, might need to be processed as a priority, leaving insufficient ramp services to process another aircraft at gate Y. Delay incurred as a result of this at gate Y would almost certainly be coded as 'turnaround', although it would arguably be more usefully attributed to the cause of the late arrival of the aircraft at gate X (which could be some other non-ATM reason, or due to ATM).

Nevertheless, it is clear that airports play a pivotal role in helping to manage delays, particularly through efficient (and flexible) turnaround processes. Airport operators have to strike a balance between the quality of service offered (such as gate availability) and declared capacities (usually constrained by runway use).

Just as putting buffers into schedules costs airlines money, having spare capacity to cope flexibly with aircraft turnarounds incurs costs at the airport. Handling agents could employ twice as many staff to better cope with disruptions caused by delays, but they would become uncompetitive in the marketplace if they did so. Airports and handling agents also have to strike a balance when managing the strategic and tactical costs of delay.

Focusing on the ATFM context, most airport ATFM delays originate from the arrival airport, although a noticeable exception is Paris Charles de Gaulle, where in 2006 approximately 15 per cent were caused by ATFM departure regulations, as reported by EUROCONTROL (2007j). As we have mentioned, weather is an important driver of ATFM delays at airports, especially at hubs operating close to capacity (Heathrow, Frankfurt, Paris Charles de Gaulle, Amsterdam and Munich – the last two of which, however, achieved noticeable reductions in weather-related ATFM delays in 2006).

At some airports, as a measure to alleviate capacity constraints, differential pricing of landing charges is applied in order to spread demand. IATA is opposed to these 'traffic distribution formulae' in principle, declaring that they are impractical in the international context of air transport. It argues that whereas such a mechanism may be helpful to one airport, that mechanism may have knock-on effects which are prejudicial to another. IATA favours, instead, the process of its Schedules Conference, which has already been discussed.

4.2 A Model for Calculating Costs of Delay

When calculating the costs of delay, it is not correct to assume that all costs are unit costs. A minute of tactical delay usually generates a marginal cost, i.e. the cost of doing something, such as an aircraft waiting at-gate, for a minute later than planned, and usually a minute longer than planned. Some of the associated costs are the same as the unit cost, an example being an en-route ATC charge. If the same route is flown by the aircraft, the charge is the same, whether the aircraft arrives at a national boundary at 0700, or 0900. Other costs are very different, depending on the time at which they occur: for example, a connecting passenger's arrival on the first leg of their journey. Arriving at 0900 is likely to incur much greater costs than arriving at the planned time of 0700. Each type of cost associated with a tactical delay has its own sensitivities to the time of occurrence and the duration. These marginal costs are, by definition, fully escapable – if no tactical delay is incurred, no corresponding marginal cost arises.

These marginal costs are in contrast to the strategic costs of managing delay, by adding buffer to schedules, for example. Strategic costs tend to be similar to, or the same as, unit costs. Figure 4.2 shows that some costs, such as aircraft depreciation, rentals and leases should be assigned to the strategic level of cost allocation, e.g. when fleet planning is undertaken as the schedule is developed. These types of cost are at best only partially escapable at the tactical level, and are essentially sunk costs at the unit rate. Such concepts are summarised in the working definitions presented in Table 4.6, which are offered to help clarify the issues pertinent to the context of this discussion, rather than as fixed economic principles.

Table 4.6   Contextual definitions of unit and marginal costs

Cost	Definition	Example
Unit	An average cost which is often fairly linear in the amount of good or service purchased, and based on a planned activity	Leasing aircraft at the strategic level of planning schedules. Associated with strategic management of anticipated delay
Marginal	An extra cost, incurred in addition to a unit cost, often non-linear in the (dis)utility and escapable in the short term	Re-booking passengers onto another flight due to missed connections. A tactical cost incurred due to actual delay

By building one model for the strategic costs of delay and another for the tactical costs of delay, for specific aircraft, in specific phases of flight, it is possible to make a quantitative estimation of the amount of strategic cost which should be invested to offset the tactical costs – for example: how much buffer to put in a schedule. Airlines have limited resources to invest in their business. The more an airline tries to anticipate delay in advance, the more likely it is able to deal with it on the day and to operate a network in a predictable way, but the money for investing in off-setting this risk strategically, and the money for paying for the consequences tactically, all comes off the bottom line.

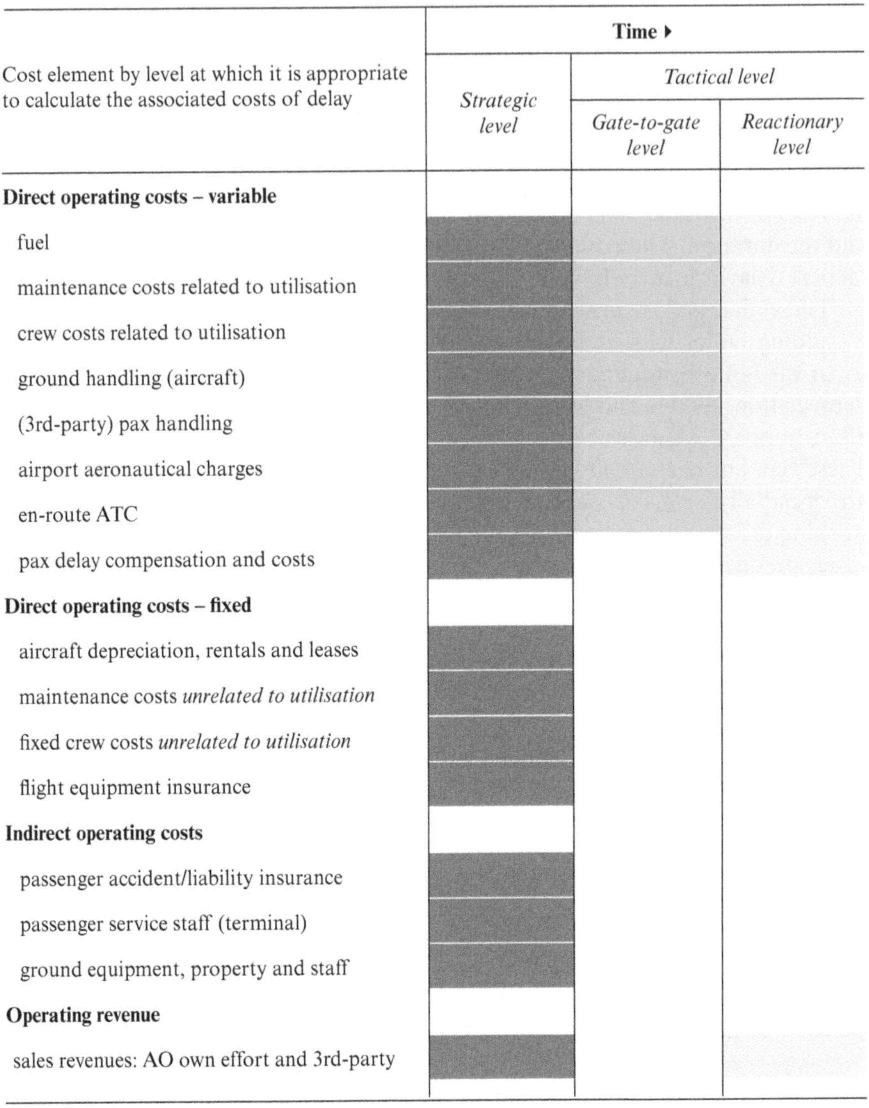

Cost element by level at which it is appropriate to calculate the associated costs of delay	Time ▶		
	Strategic level	*Tactical level*	
		Gate-to-gate level	*Reactionary level*
Direct operating costs – variable			
fuel			
maintenance costs related to utilisation			
crew costs related to utilisation			
ground handling (aircraft)			
(3rd-party) pax handling			
airport aeronautical charges			
en-route ATC			
pax delay compensation and costs			
Direct operating costs – fixed			
aircraft depreciation, rentals and leases			
maintenance costs *unrelated to utilisation*			
fixed crew costs *unrelated to utilisation*			
flight equipment insurance			
Indirect operating costs			
passenger accident/liability insurance			
passenger service staff (terminal)			
ground equipment, property and staff			
Operating revenue			
sales revenues: AO own effort and 3rd-party			

Figure 4.2   Model for delay cost elements by level of calculation

4.2.1   A Gate-to-Gate Model for Calculating Delay Costs

In a study undertaken for EUROCONTROL (Cook et al., 2004), a gate-to-gate delay cost framework was developed (see Figure 4.3), and populated for each of the appropriate phases of flight, for a cross-section of 12 aircraft (from turbo-props to widebody jets), for different cost scenarios (low, base and high: to illustrate cost ranges) and according to two delay durations.

The two delay durations were labelled as 'short' and 'long', with values of 15 and 65 minutes, respectively, chosen for the calculations to represent these types of delay. The absolute values chosen were of less importance than their order of magnitude, although specific numbers were required in the model used to produce quantitative outputs. A delay duration of 15 minutes was selected as the lower magnitude as it was anticipated that this value might incur additional airport charges (e.g. for occupying the gate), and/or may cause passengers to miss particularly tight connections, and/or may result in a crew just running out of hours. The likelihood, and values, of many such costs modelled were higher for the 65-minute delay. Some costs were fairly linear with respect to time, in the model, whilst others were essentially step-functions. Delayed passenger costs to the airline are a good example of the latter. These are discussed next.

cost allocation phase ►	direct @ ground A					direct airborne				incurred @ ground B			
OOOI sequence ►	- (IN) -		- OUT -			- OFF -				- ON -			- IN -
description ►	@ gate A		off-gate A			airborne				off-gate B			@ gate B
▼ cost element	GPU only	APU only	active taxi out	statnry ground	take-off roll	climb-out (to ToC)	en-route	arrival mngmnt	ToD to t'down	landing roll	statnry ground	active taxi in	@ gate B
fuel		[val]	[val]	[val]		---	[val]	[val]	---	---	---	---	
maintenance	[val]	[val]	[val]	[val]		---	[val]	[val]	---	---	---	---	
flight crew salaries and expenses / Cabin crew salaries and expenses	[val]	[val]	[val]	[val]		---	[val]	[val]	---	---	---	---	
depreciation of flight equipment / rental of flight equipment / amortisation of flight equipment leases	[val]	[val]	[val]	[val]		---	[val]	[val]	---	---	---	---	
flight equipment insurance	---	---	---	---		---	---	---	---	---	---	---	
station expenses (ground & pax handling)	---	---											[val]
passenger service staff (terminal) / ground equipment, property and staff	---	---	---	---			---	---		---	---	---	
airport charges (e.g. landing)	[val]	[val]			[val]					[val]			[val]
en-route & approach air navgn charges						---	---	---	---				
all other pax costs	►	►	►	►	►	►	►	►	►	►	►	►	[val]
column totals	[val]	[val]	[val]	[val]	[val]		[val]	[val]		[val]			[val]
proportion of col. total allocated to phase	0.81	0.09	0.04	0.06	1	0.2/0.7		0.8/0.3		1	---	---	1
=> average cost per minute for phase				[val]				[val]			[val]		
avg cost per min incl. incurred costs @ B				[val]				[val]					

Figure 4.3   Gate-to-gate delay cost framework

4.2.2   The Cost of Unpunctuality – Modelling the Passenger Perspective

Different researchers have adopted different methods for assigning airline costs associated with delayed passengers. Some may include the passengers' own costs which they do not pass on to the airline, even including the well known concept in transport research of passengers' Value of Time.[8] Whilst these costs certainly have a role to play in economic and behavioural modelling, in the current context it is appropriate to restrict the costs used to those actually incurred by the airline. However, it is not the case that even all of these costs will be known by the airline in terms of their impact on the bottom line.

Consider costs such as those of re-booking passengers who miss connections onto other flights, of meal vouchers, overnight accommodation and compensation payments – although all of these are straightforward to define, airlines very rarely can give any detailed breakdown of such costs. Since these are, in theory at least,

---

8   Wage rates are sometimes used as proxies for these.

readily quantifiable, and could be traced through financial transactions, these types of cost may be termed 'hard' costs.

However, some costs which impact on the airline as a result of delay are far more difficult to calculate. These are costs which are associated with market share, since a certain segment of the passenger market will be driven by punctuality. As mentioned, several traditional airlines market their brand strongly on this. The market share of their highest-yield (time-sensitive) passengers will be particularly dependent upon it in many cases. The rate of transfer of custom as a result of delay, from a less punctual airline to a more punctual one, will depend on many factors, not least journey purpose, but also comparative schedules and frequent flyer programmes (FFPs). Price is also an obvious influencing factor, although the higher-yield market is somewhat less sensitive to this and has a shorter repurchase cycle. Such transfer could be short-lived, more permanent, or even to another mode of transport (such as high speed rail, where appropriate) or behaviour (such as travel postponement or activity substitution).

These market share reduction costs may be termed 'soft' costs, due to a lack of corresponding, hard financial quantification. Of course, passengers change behaviours and loyalties for reasons other than punctuality and experiences of delays, but these soft costs make a large, and very real contribution to airline costs as a result of delay.

Suzuki (2000) offers a list of previous papers which have modelled passenger demand (using market share or sales) as a function of on-time performance and other exogenous variables, going on to build a model based on loss aversion theory, where passengers give heavier weights to losses (disutility) than to gains, although the model does not use any market research data. Although market research data are an important element of understanding such behaviour, care has to be taken in interpreting them. As Bates et al. (2001) point out: 'there is a suspicion that respondents are protesting about the unreliability of public transport services, and therefore manifesting excessive disutility from late arrival'.

In independent modelling carried out by two European carriers (one using market research data and the other a market share/cost model), both included the 'soft' costs of unpunctuality. Austrian[9] (Nichols and Kunz, 1999) and 'Airline X'[10] (in 2003) both arrived at very similar values. These were (remarkably) within 20 per cent of each other, after correcting both to comparable 2003 values based on Airclaims' fleet and AEA load-factor data. Averaging over delays of above 15 minutes, gave a value of €0.40 per such delay minute, per average passenger, per average delayed flight. Austrian further estimated[11] 60 per cent of their calculated costs to be attributable to 'soft' costs, which translates to €0.24 of the derived €0.40.

Before moving on to how these costs need to be scaled up using reactionary delay multipliers, it is worth closing the discussion of the more elusive costs associated with delay, with a final example, although one which will not be

---

9   'Austrian Airlines' at the time.
10   Series of personal communications.
11   Personal communication.

Table 4.7   Hard and soft costs of passenger delay to airlines

Type of delay cost	Definition	Calculated value (2003 base year; delays > 15 mins*)
Hard cost	Quantifiable through financial transactions alone, e.g. compensation paid	€0.16
Soft cost	Not quantifiable through financial transactions alone – needs additional market research data, e.g. on loss of market share	€0.24
Total cost	Sum of above two	€0.40

*   Per such delay minute, per average passenger, per average flight.

quantified. As touched upon earlier in the airport context, there is also the grey area of delay costs incurred by other parties, which are passed on to the airline. Although, in theory, they are soft costs, in that it would be well nigh impossible to point to them in any account or charging mechanism, they exist as hard costs, in that the airline pays for them. A good example is that of the handling agent at an airport. Although most handling agents will not charge an airline for an aircraft arriving 30 minutes late at the gate, they are used to coping with such contingencies and, as a result, will have extra capacity in order to deal with turning say ten aircraft around at midday, instead of the expected nine. The typical handling agent would indeed be hard pressed to say what fees would be charged if all delays were reduced by ten minutes. Much the same may be said for airport charges and ATC charges paid by the airlines – these costs could be lower in a perfect world, with no delays, but quantifying these is another matter.

4.2.3   *Reactionary Delay Costs – Extending the Gate-to-Gate Model*

As detailed in Table 4.5, these are the second largest cause of all departure delays greater than 15 minutes, running a close second to turnaround delays. This is a complex issue, however, as recording a delay as 'reactionary' is not always as informative as might be desired. If an aircraft arrives 30 minutes late inbound at the gate, then leaves 45 minutes late on the next outbound leg, it is not unknown for the (whole) 45-minute departure delay to be recorded as 'reactionary'. Moreover, even describing 30 minutes of delay as 'reactionary' is not that helpful. An aircraft could be held up on its first departure of the day by a technical fault, then run 30 minutes late for the rest of the day, causing a whole series of reactionary delays to be recorded. It would be more instructive to attribute and apportion these back to the original cause(s), and subsequent contributions, by extending the currently somewhat limited IATA reactionary delay codes into a more comprehensive system. Several airlines already do this for internal analysis, and apportion the delays in a more disaggregate manner.

A model was built by Beatty et al. (1998) in which delay propagation was studied using actual American Airlines' schedule data, looking at specific delays to specific flights, and tabulating the delay multipliers. Taking one example, the model showed that for a primary delay of 7.5 minutes at 0615, the basic reactionary delay multiplier is 1.21, i.e. a further 1.6 minutes of delay were generated in the network.

Averaged over the whole European network, EUROCONTROL (2007j) states that every minute of primary delay resulted in 0.54 minutes of reactionary delay in 2003, and 0.76 minutes in 2006, on average. These values may be expressed as basic reactionary delay multipliers of 1.54 and 1.76, respectively.

This is not far from the Bratu and Barnhart (2004) passenger delay ratio of 1.7, discussed earlier, although the values are comparable only in a rather loose sense, because of the rather different basis of the calculation of each. An application of a very specifically-derived ratio such as this in other models would have to be assessed in terms of the way net (or average) costs were re-distributed across the network and how the impacts of cancellation were incorporated. Clearly, multipliers cannot be used in a way which allows the originally-derived total cost to be exceeded, so temporal costs such as the €0.40 per minute referred to earlier could not be used with incompatible multipliers.

Returning to the study undertaken for EUROCONTROL (Cook et al., 2004), the issue of reactionary delay multipliers was also explored here in some detail. For 'short' delays, a basic reactionary delay multiplier of 1.20 was calculated, and for 'long' delays, a value of 1.80: these values were thus consistent with the *network average* of 1.54 for 2003 cited by EUROCONTROL, and were calculated in the same year.

To avoid double-counting of reactionary delay costs to the same aircraft (rotational reactionary delay) it is necessary to differentiate these from reactionary delay to other aircraft (non-rotational reactionary delay) – see Table 4.2. This latter type of delay may be assumed to occur almost exclusively at-gate, as delayed pushbacks of other aircraft. The study (ibid.) estimated these to be 25 per cent of the total reactionary delay minutes generated, based on EUROCONTROL data. This produced (scaled down) non-rotational reactionary delay multipliers of 1.05 and 1.20, which were used as cost scalars for the per-minute costs of primary delay ('short' and 'long', respectively). The 'other' aircraft (delayed at-gate) were assumed to be aircraft of the same type, to simplify the calculation. In reality, it is probably more likely that aircraft, on average, feed larger aircraft.

Since the cost of *passenger* delay to the airlines quoted earlier was calculated per average delay minute, per average delayed flight, the basic reactionary delay multipliers of 1.20 and 1.80 are appropriate as they stand, since they are effectively *averaged* over the whole network: such passengers are oblivious to whether the delay is primary or reactionary. Table 4.8 summarises the calculation for the 2003 base year, and makes a crude estimate for 2007, assuming the 2006 value cited by EUROCONTROL gets no worse and that the 25 per cent adjustment for the non-rotational case remains valid.

In some respects, these ('long' delay) values might be underestimates. As delay effects escalate and multiply through a network, they become more expensive,

Table 4.8   Reactionary delay multipliers for 'long' delay types only

Type of ('long') delay multiplier	Used for multiplying	Value estimated in 2003	Crude value estimated* for 2007
Basic reactionary	Passenger costs of delay to airline	1.80	2.06
Non-rotational reactionary	Non-passenger costs of delay to airline	1.20	1.26

*   Basic reactionary = 1.80 x (1.76/1.54) = 2.057.
      Non-rotational reactionary = 1 + (2.057 – 1.00) x 25% = 1.26.

as the network becomes further and further from its planned operational state, although the model does not really account for this. Furthermore, Beatty et al. (1998) point out that:

> It is also understood that airlines react to large delays by cancelling flights and reassigning resource to minimize delay propagation. These reactions are also costly to the airline as resources are de-optimized and passenger revenue is lost. So, while it may be difficult or impossible to calculate these costs, it is possible to use the cost calculated by [delay multipliers] as a conservative surrogate.

The rise in the value of the basic reactionary delay multiplier from 2003 to 2006 reflects, at least in part, an increased sensitivity of airline operations to primary delay, i.e. a reduced ability to cope with it. This is likely to be caused to some considerable extent by increased pressures on aircraft utilisation (and, indeed airport infrastructure, as EUROCONTROL points out). Buffer reduction is further contributed to by the particular growth of the low-cost market in 2006, as noted above, although such multiplier values will vary from carrier to carrier, depending on the type of network operated, and the degree of buffering used.

4.2.4   *Hard Costs and Using Them in Hard Decisions*

Table 4.9 shows some costs calculated for the different delay durations discussed and with the reactionary effects detailed in the previous section also included.
      Several observations are immediately apparent from the table. For short delays (not exceeding 15 minutes), at-gate delay costs are very low. Indeed, of airlines who carry out any form of delay costing and re-route[12] trade-off calculations, many don't allocate any cost to the first few minutes of at-gate delay. The at-gate costs shown assume the aircraft's auxiliary power unit (APU) and engines are off. Averaged at-gate costs, with some APU use, are a little higher. As would be expected, minutes of delay which are part of long delays are much higher than those which are part of short delays. Care must be taken in using these figures,

---

12   See Chapter 2.

Table 4.9   Costs of delay (2003 base) by phase of flight and aircraft type

Aircraft type and number of seats		'Short' delay costs: € min⁻¹			'Long' delay costs: € min⁻¹		
		Ground	*Airborne*		*Ground*	*Airborne*	
		At-gate	*En-route*	*TMA*[b] *arrival*[c]	*At-gate*	*En-route*	*TMA*[b] *arrival*[c]
B737-300	125	0.9	13.7	15.2	74.4	87.2	88.6
B737-400	143	1.0	13.6	14.5	84.4	97.1	97.9
B737-500	100	0.9	12.5	14.0	62.7	74.3	75.8
B737-800	174	0.9	14.2	12.1	99.4	112.7	110.7
B757-200	218	1.0	18.4	15.6	122.5	139.9	137.1
B767-300[a]	240	1.2	26.0	21.7	142.2	167.0	162.6
B747-400	406	2.3	52.3	39.8	238.8	289.1	276.6
A319	126	1.0	13.1	10.7	75.2	87.3	84.9
A320	155	0.9	13.2	11.7	90.1	102.4	100.9
A321	166	1.0	15.7	14.7	95.4	110.1	109.1
ATR42	46	0.6	2.8	2.6	31.3	33.6	33.3
ATR72	64	0.7	4.0	3.3	40.8	44.1	43.4

a   B767-300ER.
b   Terminal Manoeuvring Area (see Chapter 1).
c   I.e. arrival management, such as holding.

however, as delays accumulate: if an aircraft has unrecovered delays of 15 minutes on each of three consecutive legs, it will be running 45 minutes late relative to schedule on the last leg, thus incurring the types of cost on the right-hand side of the table. As might be expected, 'airborne delays' (such as en-route extensions or during arrival holding) are more costly than holding on the ground, as these consume fuel. Thus, ground holding is more cost effective (and produces less CO_2) than airborne delay management, although aircraft cannot be held indefinitely, waiting for 'optimal' flow conditions: a balance has to be struck.

Indeed, by comparing these airborne and at-gate costs, it is possible to estimate trade-offs between a late take-off slot, or an earlier slot with a longer route, i.e. more airborne time. On average, flights with 'short' ATFM slot delays (e.g. up to 15 minutes) are not usually worth re-routing, whereas for longer delays, an airborne extension of E_A minutes is typically worth accepting if it reduces the ATFM slot by around $[(1.1–1.3)E_A + 10]$ minutes (if at least some passenger delay and crew overtime costs will be incurred by the airline). Thus, a re-route with 15 minutes of extended airborne time is worth accepting if it brings a slot forward 27 minutes closer to schedule.

Strategic costs of delay, as has been discussed, can be encountered in a number of different contexts. A common one is the incorporation of buffers into schedules.

By allocating Euro values to such strategic costs, and comparing them to the tactical costs shown in Table 4.9, it is possible to estimate the amount of buffer which it makes economic sense to add to a schedule. It is noteworthy that these strategic costs should also be considered at the *network* planning level (i.e. to off-set tactical, *reactionary* costs as well) and that, in marked contrast to tactical costs of delay, each minute of strategic buffer costs roughly the same, since they are basically unit costs.

It has already been mentioned that added buffer minutes are often based on a historical knowledge of past delays, and may be added to the at-gate time or airborne phase. The former is cheaper, but accurate anticipation of the latter helps with better fuel and maintenance planning. Since adding buffers to schedules is a strategic level activity, even allowing for some limited cost recoveries for unused buffer minutes, they are still rather costly. Just five minutes of unused buffer, at-gate, for a B767-300ER, would amount to well over €50 000 over a period of one year, on just one leg per day.

Considering 'long' tactical delays at the gate, the strategic costs per minute are in a fairly narrow range of around 25–35 per cent of the corresponding tactical costs. Taking the specific example of a B737-300, the costs imply a rule of thumb that if more than 22 per cent of flights have an expected delay of more than 15 minutes, it should be cost effective to use a number of buffer minutes equal to the average tactical delay.

Further operational factors also have to be considered. The airline may be constrained in other ways (such as the slack time in Table 4.3) or by having to 'risk' a lower amount of buffer in order to avoid limiting the number of rotations in the day: profits, and market share, are also a function of frequency, as has been discussed in Chapter 6. It may also be decided to put relatively more buffer than is statistically needed into schedules earlier in the day, and to risk lower buffers later on, when the impact of tactical delay is less severe, as mentioned.

It is also apparent that tactical delay costs are themselves dependent on the amount of buffer added to schedules. If no buffers were used, tactical costs of delay would increase markedly. In particular, the reactionary delay multipliers of Table 4.8 would be significantly larger. The values used in the calculations in this chapter are implicitly based on the current equilibrium, and such results should not be extrapolated too far outside of this environment.

Examining these types of costs, both within and between the strategic and tactical levels, informs airlines' decision making in other tactical areas, such as aircraft and crew swaps, when to cancel or converge flights and when to hold connections or to re-book passengers instead.

Such quantifiable trade-offs, even within reasonably generic and indicative frameworks, may also be used to help airspace designers and flow managers, plus ATM planners, to better design airspace structures and traffic flows in terms of the cost implications to the airlines. As we have noted, actual costs will vary not only from airline to airline, but even from flight to flight, and depend on the network and infrastructural context in which they arise.

4.2.5   Cost Evolution – Where Next?

Taking selected values from Table 4.9, for the ubiquitous A320 and B737-300, it is possible to crudely estimate 2007 'long' delay costs from these earlier values, which were published in 2004, but with a 2003 cost base. In Table 4.10 and Table 4.11 the values are updated using a mixture of known data and informed estimates. The reactionary multiplier values derived in Table 4.8 for 2007 are used (although it is perhaps worth reminding ourselves that this does not mean that the costs are simply multiplied by these reactionary multipliers – their proper use has been described above). The costs have been simplified into four categories. 'All other costs' includes items such as maintenance costs and airport charges: see Figure 4.3.

Table 4.10   Estimated 2007 'long' delay costs for A320

Cost component	2007 % increase on 2003	2007 reactionary multiplier	At-gate		En-route	
			2003 € min⁻¹	*2007 € min⁻¹*	*2003 € min⁻¹*	*2007 € min⁻¹*
Fuel cost	135%	1.26	0.0	0.0	12.2	28.6
Pax hard cost	15%	2.06	31.1	40.9	31.1	40.9
Pax soft cost	0%	2.06	46.7	53.4	46.7	53.4
All other costs	8.5%	1.26	12.3	14.0	12.4	14.2
Total	n/a	n/a	90.1	108.3	102.4	137.1

Table 4.11   Estimated 2007 'long' delay costs for B737-300

Cost component	2007 % increase on 2003	2007 reactionary multiplier	At-gate		En-route	
			2003 € min⁻¹	*2007 € min⁻¹*	*2003 € min⁻¹*	*2007 € min⁻¹*
Fuel cost	135%	1.26	0.0	0.0	12.6	29.6
Pax hard cost	15%	2.06	25.1	33.0	25.1	33.0
Pax soft cost	0%	2.06	37.6	43.0	37.5	43.0
All other costs	8.5%	1.26	11.7	13.3	12.0	13.6
Total	n/a	n/a	74.4	89.3	87.2	119.2

Taking fuel prices first, these have increased approximately[13] 135 per cent on the values used in Table 4.9, and this substantial rise has been used in Table

---

13   Sources: IATA Jet Fuel Price Monitor (2007); US Energy Information Administration (2007).

4.10 and Table 4.11, without any attempted correction for changes in engine or route efficiency.

Turning to the passenger costs, a potentially major impact since 2003 was the introduction of Regulation (EC) 261/2004, which came into force on 17 February 2005. The Regulation only relates to departure delay (nothing is actually due to the passenger for any type of arrival delay or missed connection, *per se*), denied boarding, and cancellation. It confers passengers with rights only at the point of departure. In a recent review of the potential impacts of the Regulation, Jovanović (2006) states that this may have increased certain airline costs in this specific category by as much as 30 per cent[14] in some cases, although many airlines have probably managed to escape a significant cost burden in practice, for example by focusing (more) on re-booking on their own networks (or those of alliance partners, if necessary), and avoiding the need to compensate wherever possible. Jovanović also suggests that very few changes in operational practice (such as increasing buffers) have occurred as a result of the Regulation, and although further remarks that Lufthansa's Annual Report (2005) states that the Regulation 'could lead to additional annual costs totalling up to €10m per year', there was no further reporting on this in the 2006 Annual Report.

From EUROSTAT data, compounded Euro area inflation[15] for the whole period 2003–2007 is around 8.5 per cent (and this value has been applied to the 'all other costs' category). It seems reasonable to suggest that the impact of Regulation (EC) 261/2004 has reached 30 per cent for rather few carriers, as increasing financial pressures on airlines forces ever greater economies where possible, particularly to off-set severe cost increases in fuel, which are not in their control. Tables 4.10 and 4.11 use estimated values of 15 per cent, i.e. half the cited value of 30 per cent, but still in the approximate region of twice the inflationary value.

As for soft costs for passenger delay, no increase on the 2003 value has been used. As discussed in Chapter 6, this reflects the argument that the market has become increasingly cost driven – some 'traditional' airlines no longer provide free catering on shorter hauls, and we may refer again to the considerable growth in the low-cost market in 2006, which also carries considerable business-purpose traffic.

The main impact of these estimated changes is that the fuel cost has caught up somewhat with the passenger components, meaning that re-routes involving greater fuel burn are proportionately less attractive, and future environmental charging mechanisms (see Chapter 5) are only likely to increase this effect. The

---

14   Example cited excluded re-booking, such that overall changes to the total hard costs in this category could be higher or lower, depending on the change in re-booking costs. Deepening alliance structures are likely to help control these.

15   'Euro area' data sourced from EUROSTAT. 2003–2006 based on annual values; 2007 value based on May 2007. As defined by EUROSTAT: 'Euro area inflation is measured by the MUICP ("Monetary Union Index of Consumer Prices" as defined in Council Regulation (EC) No 2494/95 of 23 October 1995) which is the official euro area aggregate … New Member States are integrated into the MUICP using a chain index formula'.

same may be said of the airlines' ability to cope with delayed passengers somewhat more cost effectively through alliance networks. The latter does not apply to low-cost carriers, and is also negatively impacted by increasing load factors.

Also off-setting such effects, to some extent, is the increased use of larger aircraft, for example at airports where slots are at a premium (as discussed in Chapter 6), combined with generic improvements in aircraft fuel efficiencies.

4.3   Conclusions

A question often asked about delay cost management is why airlines have not already carried out more evaluations of these enormous costs themselves, and why so relatively few studies have attempted to address the issue. Part of the problem lies with the fact that airlines are already operating under intense operational and financial pressures, without the additional burden of commissioning or carrying out this type of research. There are exceptions, but they are very few and far between. A serious barrier to this type of calculation is the need to integrate so many disparate pieces of financial and operational data into a cohesive picture, and such an effort would take up considerable resources for a significant period of time. It is also exceptionally rare for carriers to track specific costs of delay. Airport charges are invariably all lumped together – no separate allocation is made for delayed aircraft. Some airlines collate figures about additional fuel burn due to route inefficiencies and delays. Several direct passenger costs are easier to extract from accounts. The list of data requirements is a long one, however.

There are two fundamental challenges here. One is the development of better tools and resources to help manage the costs of delay. The other is the actual reduction of delays *per se*. The latter has the longer lead time but, meanwhile, the former can be developed to tackle the problem more immediately and to inform on-going processes of change and (hoped for) amelioration, for example under the SES initiative and SESAR.

Such tools will need to furnish a better integration of dynamic flight management with fuller cost data (continuing a move away from fuel-only models), to the extent whereby there is a more widespread use of tactical communication between the airline back office and the aircraft, for example between the GDS and the FMS. Critical challenges here are to improve the understanding of passenger costs, particularly soft costs (market research could have a strong part to play in this), to better consider the cost of cancellations, and to effectively integrate ATM into the whole process.

In terms of reducing delays, we will see in Chapter 5 that the benefits offered by communication, navigation, surveillance and air traffic management (CNS/ATM) measures are key to future, improved efficiencies, allowing more direct routings, and to reducing delays. Decreasing fragmentation and increasing integration (see Chapter 7) are also necessary steps on the path towards such progress. EUROCONTROL (2006a) presents an illuminating statement, in that intra-European routes are significantly less efficient than domestic ones: 'If the European route network was as efficient as the domestic networks, as one would expect under the Single European Sky, €150 million to €300 million could be saved every year'. Such improvements are necessary if true 4D (time managed) navigation is to be realised. Existing operational practices such as Required Times of Arrival, and emerging developments such as slot swapping, will support concepts in development such as EUROCONTROL's 'flight prioritisation concept' for weighting aircraft movements by factors including the number of connecting passengers on board or a potential curfew infringement. The key here is to also achieve more continuous flows into capacity constrained parts of the infrastructure, such as airports, but at rates which may be properly supported – and to therefore improve the collaboration between stakeholders, on a gate-to-gate basis. These are difficult challenges, but they are not impossible.

All stakeholders, from airlines, airport authorities and handling agents, to ATC and ATM, need to strike a balance between strategic and tactical delay cost management. ATM is certainly not the only cause of delay, and airlines are not alone in suffering the costs. These balances are made more difficult to achieve by the lack of transparent cost data, and this extends to understanding the cost of capacity. Air traffic management relies on flexibility to cope with uncertainty. This flexibility both contributes to delays and can be a future tool for reducing them.

Chapter 5

European ATM and the Environment

Victoria Williams
Imperial College, London

5.1   Introduction

In earlier chapters of this book, we have touched upon the very large growth in air transport since the middle of the twentieth century, focusing on how ATC and airlines have managed this change. This chapter will focus on the environmental consequences of such growth, before we explore the evolution of air transport, its infrastructure, and the societal consequences, in following chapters.

Aviation's impacts have proved difficult to quantify and control and international operations present special challenges for policy makers. There have been some successes. Advances in aircraft and engine technology have dramatically improved fuel efficiency since the introduction of the first jet engines. International guidance and regulation through the International Civil Aviation Organisation (ICAO) has reduced both aircraft noise and the emissions during landing, take-off and taxiing. Nevertheless, growth in air travel has led to an increase in total environmental impacts, both locally and globally.

This chapter focuses on three environmental issues associated with aviation: noise, air quality and climate change. Each issue is discussed in the context of the way it constrains the European air traffic management system and of the achievements and opportunities in the mitigation of impacts. The chapter also reviews the problem of comparing and prioritising environmental impacts and discusses the synergies and feedbacks between impact mitigation options.

5.2   Noise

Exposure to noise affects sleep, with consequences for health and well-being. It can also affect concentration and communication, affecting homes and workplaces. Where a school is exposed to noise nuisance, it can have a detrimental impact on the performance of students (Haines et al., 2001).

Perceptions of noise nuisance also vary significantly between people and according to circumstance. Tranquillity is a growing issue, as the environment strongly influences how loud (and how unwelcome) a noise is perceived to be.

Aircraft noise is an important issue for communities near airports and under flight paths. It is the environmental impact of air travel that attracts the most complaints from the public. Many of its effects are readily apparent and (at least partly) immediate. Concerns about the impact on property prices can exacerbate potential noise nuisance and further contribute to public complaints. The prevalence of complaints may also be due in part to public perceptions of the feasibility of change; while reducing an aircraft's emissions may be perceived to be a complex technological issue, scheduling or routing changes to limit noise nuisance can be seen as realistic options and ones which complaint letters could potentially influence (the societal context of ATM change will be explored further, in Chapter 8). In contrast, are the hidden impacts of emissions, where the relative contributions of aviation and other sources may be poorly understood by many.

5.2.1   Sources of Noise from Aviation

At airports, noise is generated by aircraft operations, ground support vehicles and infrastructure operations. Noise from access traffic (both passengers and staff) will also contribute, but aircraft noise dominates public and political concerns. During take-off and climb aircraft noise comes mainly from the engine; for approach and landing the airframe is also a significant source. In addition to jet engines, aircraft auxiliary power units (APUs) contribute to airport noise and airports may limit their use, either by setting time restrictions or by offering an alternative power supply for aircraft support systems on the ground. Suggestions for radical aircraft redesign to reduce aircraft noise include adopting a blended-wing body and embedding engines into the aircraft in order to shield the ground from noise (Dowling and Hynes, 2006).

5.2.2   Measuring Noise Exposure

Objective measures of aircraft noise exposure are crucial for the monitoring and control of noise. The selection of measures will influence the identification of noise problems and their possible solutions. Sound intensity (the rate at which energy passes through a unit area) is generally described using the Sound Intensity Level measure (in decibels, dB) which reflects sound intensity relative to the threshold of audibility by humans. Since aviation noise is intermittent, the exposure level for comparison with other sources such as roads or industrial sites is measured using equivalence. It is defined as the level of steady sound that, over the period of measurement, would deliver the same noise energy as the actual intermittent noise.

The period used for calculating equivalent exposure can vary. In the US, the Federal Aviation Authority requires the use of day-night average sound level (DNL) for assessment of noise exposure. This includes a weighting for sounds between 2200 and 0700, to reflect the increased perceived nuisance of night noise exposure. The European Environmental Noise Directive applies an adapted approach (Lden) which represents an average day, giving additional weight to evening and night-time noise by adjusting by +5dB for flights during a four-hour evening period, and by +10dB for flights during an eight-hour night period. UK calculations follow an alternative approach, using a 16-hour averaged equivalent noise value (Leq), as alternative processes are used to constrain night noise.

These assessments of noise exposure relate to specific locations. Measures for the airport as a whole are given by the area and population enclosed by specific DNL, Lden or Leq contours. Population distribution and changes in contour shape mean that reductions in the area measure may not be linked to changes in the population measure.

In the past 15 years, both the population and area enclosed by the 57 dB Leq contour around London's Heathrow airport have decreased, despite an increase in traffic movements. In fact, the population has decreased faster than the area enclosed (Civil Aviation Authority, 2005a). Heathrow is not representative, however; it is capacity constrained and the increase in traffic has not been as rapid as at other airports. At London Stansted, the increase in traffic movements in this period has been fourfold. The population enclosed by the 57 dB Leq contour (Civil Aviation Authority, 2005b) is double that in the late 1980s and early 1990s, but less than half that seen at the peak, in 1998. Across Europe, noise exposure contours have undergone changes due to changes in aircraft mix, of land use and in the number and type of aircraft operations.

Since average noise equivalence is used, quieter individual aircraft movements can allow an increase in the number of flights without increasing the area within a given noise level contour. At Stansted, although the population within the contour doubled, the area enclosed by the contour was the same in 2004 as in 1988, despite the fourfold increase in traffic movements.

Current noise metrics may not fully reflect the health and social impacts of noise and the nuisance of noise exposure. Certainly, there are assessment limitations, particularly with respect to the distribution of flights throughout the day – does an overflight every six minutes for 16 hours cause more or less nuisance than an overflight every 90 seconds for four hours? The response to such questions is subjective and influenced by a wide range of factors. Although Leq has shown strong positive correlation with the percentage of the population reporting the level of aircraft noise to be unacceptable (Brooker, 2004), the sensitivity to disturbance has grown despite reductions in engine and airframe noise. This has been linked to a change in the balance between the severity of disturbance and the frequency in determining the annoyance felt and to declining tolerance of nuisance[1] as affluence increases (Thomas and Lever, 2003).

5.2.3   *Noise Mitigation: Achievements and Opportunities*

Annex 16 to the Convention on International Civil Aviation (also known as the 'Chicago Convention', as introduced in Chapter 1) sets a framework for the noise classification of aircraft. The classifications reflect the size of the aircraft and the engine type and are linked to a reduction in noise as technology evolves. There are no global regulations on aircraft noise; at the local level noise restrictions are often reflected in limits on the airport's licence to operate (including limits

---

1   A theme which will be broadened in Chapter 8.

on the number of movements or the total permitted noise) or on permission to expand airport infrastructure. The Annex 16 classifications are internationally recognised for use in regulation at the state or airport level. These regulations have become progressively stricter, with attention first paid to the very early (non-noise certificated) jet aircraft, then to reducing 'Chapter Two' aircraft or fitting 'hush-kits' to reduce their noise. ICAO provided a recommended structure for phasing out Chapter Two aircraft in the late 1990s. This structure recommended an initial grace period for aircraft less than 25 years old. For new aircraft, for certification after January 2006, more stringent 'Chapter Four' conditions apply.

Recently, the focus has fallen on controlling noise from the loudest of the next generation of aircraft, referred to as 'Chapter Three'. ICAO is recommending a balanced approach for noise management, considering aircraft noise as one of a suite of measures including land use management and operational procedures and restrictions. The balanced approach also recognises a strong need to take into account the special circumstances of operators from developing countries. Noise restrictions on aircraft types permitted to land can influence competition between airlines on emerging international routes. One example occurred before the entrance of new member states to the European Union in 2004. Many airlines in the accession states were forced to reduce their fleet size to make the transition to quieter aircraft and so were less able to compete on new routes (Thomas and Lever, 2003).

Airports can also impose differential landing charges on aircraft to reflect differences in the noise generated. ICAO recommends that these charges are used only at airports with specific noise problems. They also suggest that the charges should only cover costs appropriate to the alleviation of the problem (such as increased investment in insulation to reduce noise exposure in local homes, businesses and public buildings) and that charges should not distort competition or prohibit the use of particular aircraft types. (It is also to be noted that some airports also operate a practice of 'peak pricing', in an attempt to smooth out slot demand during the day, although this has not generally been successful.)

Additional procedures may be used to regulate night noise. The UK uses a system of night noise quotas. Each aircraft and engine configuration is allocated a quota count. A new, quieter quota count of 0.25 was added to the set of 0.5, 1, 2, 4, 8, 16 or 'unclassified', in 2006. These allocations are very detailed and take into account any issues affecting the noise produced, so similar aircraft models may be allocated very different quota counts. An aircraft may also have different quota count values for arrival and for departure. For example, most configurations of Boeing 747-100 aircraft have a departure quota count of 8 or 16 and an arrival count of 4. The new 0.25 quota count category includes some configurations of the Boeing 737 and the Airbus A319 and A321. Aircraft with a quota count of 8 or 16 cannot be scheduled to take off or land after 2300 and before 0700. These aircraft can take off before 2330, if the delay is not due to the operator.

Under the night noise quota system, each airport has a cap on the total value of quota points relating to operations between 2330 and 0600 over a set period. Two periods are defined for each year. Currently, there is also a cap on the maximum number of night-time operations for each airport, during each period.

There are exemptions from the quota count system for night-flight operations in emergencies or where the operations may cause, or be caused by, major system-wide delays. The arrival before 0600 of aircraft scheduled to arrive after 0630 is also exempt, unless the aircraft has a quota count of 8 or 16.

Many airports across Europe impose some restriction on night flights, with restrictions more likely to limit aircraft types and/or measures of total noise than the number of air traffic movements (Airports Council International Europe, 2004a).

In addition to the charge and quota approaches described, and to quieter aircraft technologies, noise exposure can be addressed through measures including land use change in the vicinity of airports (to reduce population exposed) or even relocation of the airport away from densely populated areas (such as the case of Hong Kong). Other options include subsidised insulation (double-glazing, loft insulation, cavity wall insulation or other measures) for local properties. In addition, there are noise reduction opportunities in the management of air traffic.

Air traffic management procedures for take-off and departure can influence noise exposure. Options are restricted by the configuration of available runways (usually a product of the prevailing wind and the geography of the airport site), but there are opportunities to reduce noise exposure, including:

(i) Continuous Descent Approach (CDA) A conventional stepped approach includes long segments of level flight at low altitude. The UK definition of a CDA is that in the phase of flight below 6000 ft, there should be no segment of level flight longer than 2 NM. There is no wider, formal definition. This flight profile reduces noise both by reducing the overall thrust required during descent, and by keeping the aircraft higher for longer. It is used at many airports, particularly for night operations, and a number of European trials are currently taking place. As well as reducing noise, emissions and fuel burn are also significantly reduced, the latter a particular motivation for the airlines. Where CDAs are described in procedures, they are normally specified from levels well below Top of Descent, although the savings from this height can be several hundred kilos of fuel, compared to a conventional, or stepped descent. A disadvantage cited by controllers is the potential reduction in capacity caused by having too many descending aircraft on CDAs at any given time. (CDAs are discussed further in Chapter 8).

(ii) Noise preferential routings (NPRs) There are large benefits in identifying and enforcing noise preferential routes, which avoid low altitude flights over highly populated areas. However, some existing routes are not compatible with all aircraft types. This is particularly problematic for old routes now flown by newer, faster aircraft. Most of the larger airports in Europe operate NPRs, and the evolution of P-RNAV (see Chapter 1) will doubtless enhance the flexibility of this.

(iii) Cross wind/tail wind landing Where there is a preferred runway direction to avoid overflying densely populated areas, aircraft may approach in non-optimal

wind conditions. Although a headwind is preferred, aircraft may land in a tailwind direction if the wind is at a low speed. In high wind situations, safety could be compromised, particularly due to the unpredictability of local wind on the runway and instability caused by gusting.

(iv) Reduced thrust The aircraft thrust settings during landing govern the size of the noise footprint. A 'low power/low drag' approach minimises the deployment of flaps before lowering the undercarriage. The practice is well established and there is little scope for expansion. For take-off and climb out, reduced thrust has additional benefits for engine lifetime and fuel burn and is currently chosen by some airlines. ICAO regulations restrict its use to above 800 ft.

(v) Displaced threshold This is a noise displacement measure only. The noise footprint is repositioned by use of alternative sections of the runway, utilising the shorter runway requirements of modern aircraft at airports with older, long runways. The assumption is that much of the noise footprint will be moved over the airport site. Any benefit will depend on the distribution of buildings around the runway, particularly with respect to side noise. There is a safety compromise in that the buffer zone beyond the runway end is lost. Additional taxiing costs (including noise and emissions) may also be incurred, depending on the configuration of the airport site. Displaced thresholds also require changes to navigation systems for automated landings (see Chapter 1) and may require system duplication to maintain options for alternative threshold use.

(vi) Steeper approach A steeper approach angle could reduce the size of the noise footprint. The normal approach angle is 3°. Current regulations require an angle no greater than 3.5° unless required for obstacle avoidance (e.g. at London City Airport, where the approach angle is 5.5°). Increasing the approach angle would increase pilot workload and require revised definitions of approach routes. Restrictions on approach angle also restrict the aircraft types that may be used.

5.3   Air Quality

Higher pollution levels in the immediate vicinity of an airport (both from aircraft and airport surface transport, including access modes) mean that health problems associated with increased levels of pollutants are possible. The main health impacts linked with poor air quality relate to respiratory complaints like asthma. Symptoms may become more frequent or more severe during episodes of poor air quality. Poor air quality can also affect vegetation and ecosystems.

5.3.1   *Aviation and Air Quality*

As the focus of this book is on air traffic management, the discussion presented here will focus on *aircraft* effects on local air quality, rather than the impacts of supporting infrastructure and access traffic, although these impacts are significant.

Aircraft engines produce several key combustion products, with different characteristics and impacts. A perfect combustion process of a hydrocarbon fuel in the atmosphere would result in outputs of carbon dioxide and water vapour. The additional presence of sulphur in aircraft kerosene adds sulphur dioxide to the list of products. However, combustion processes are not perfect and other emissions occur. These include nitrogen dioxide (the nitrogen atom in this molecule coming from the air), carbon monoxide, other oxides of sulphur, soot and unburned hydrocarbons. Emissions of each species will depend on the engine, the fuel and the mode of operation. Emissions of carbon dioxide and water vapour are largest, and are a direct function of the amount of fuel burned, so the emission index (g emitted per kg of fuel) is the same for each phase of flight. The emission rate (g per second) is highest for take-off, when the engine operates at maximum thrust.

Emission indices for carbon monoxide and hydrocarbons are highest when the engine is idling, while the engine operates at a small fraction of its maximum thrust. NO_x (NO and NO_2) emissions are highest in the high-thrust take-off phase and tend to be higher in this phase for short-haul aircraft than for long-haul, though short-haul aircraft have a lower emission index at cruise (which affects the climate impact of the flight).

Particulate matter emitted in the combustion process can also contribute to human health impacts relating to poor air quality. Advances in engine technology have reduced particles per kilogramme of fuel considerably.

Emissions from aviation combine with those from other transport modes and from stationary sources. For example, nitrogen dioxide, which can irritate lungs and reduce resistance to illnesses like bronchitis and pneumonia, is linked to urbanisation and road traffic. Concentrations are high in urban areas and along major roads. Maps for London show two clear peaks, one in central London, the second at Heathrow airport, both of which exceed the objectives set out in the Government's Air Quality Strategy. Taking the UK as an example, there have been changes in NO_x in recent years. Emissions from aviation have increased despite regulation of new aircraft engines because of the continuing growth in air travel. Between 1990 and 2004, emissions from aviation (international and domestic) rose from 95 to 169 kilotonnes. In the same period, rail emissions rose from 14 to 19 kilotonnes. Total transport NO_x emissions remain dominated by road vehicles, despite a fall from over 1500 kilotonnes in 1990, to below 600 kilotonnes, in 2004 (National Atmospheric Emission Inventory, 2006). This fall was driven by the introduction of catalytic converters.

5.3.2   *Air Quality Impact Mitigation: Achievements and Opportunities*

ICAO provides certification standards for emissions of a range of species. These limits apply to new engines, so changes associated with long-term use and maintenance practices are not reflected. The standards refer to total emissions over a standardised landing-take-off (LTO) cycle, which describes aircraft movements

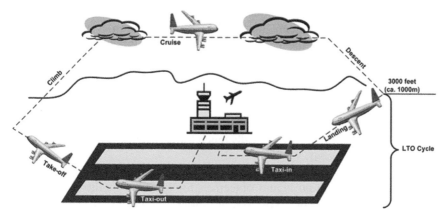

Figure 5.1   Landing-take-off cycle defined for regulating low altitude emissions

Source: Tamara Pejovic, adapted from Rypdal (2000).

below 3000 ft (including taxiing) (Figure 5.1), assuming fixed times in mode and engine outputs (Table 5.1). This allows different engine types to be compared. For the specification of standards, total emissions are divided by take-off thrust, to reflect differences in aircraft size and power.

Table 5.1   Defining the standardised landing-take-off cycle

Mode	Time	Rated output
Take-off	0.7 mins	100%
Climb out	2.2 mins	85%
Approach	4.0 mins	30%
Idle	26 mins	7%

For NO_x emissions, standards have become progressively stricter, with the most recent differentiated by both engine thrust and engine pressure ratio. Improvements in NO_x emissions are usually achieved through introductions of new aircraft to the fleet and retirement of older aircraft, rather than adaptations of aircraft in operation. Reducing NO_x presents a particular challenge as conventional approaches to improve fuel efficiency have increased NO_x, so the two environmental priorities must be balanced in engine design.

In addition to standards for gaseous emissions, there are restrictions on smoke number. This is a dimensionless number calculated from the reflectance of a paper filter before and after the exhaust gas passes through it. Collectively, these regulations have provided an incentive for the inclusion of emission reduction technologies in new engine designs.

The formation of sulphur dioxide, which can aggravate respiratory illnesses and contributes to 'acid rain', damaging buildings and vegetation, is proportional to the sulphur in the fuel, so the fuel, rather than the engine, is regulated to constrain sulphur dioxide emissions. Fuel sulphur content is governed by the sulphur in the supply of crude oil and by the use of hydroprocessing. Hydroprocessing reduces sulphur content in the fuel, but at the expense of additional CO_2 emissions in the production process.

The sulphur content in aviation kerosene is required to be <0.3 per cent. In practice, sulphur contents are typically much lower, with the global mean in the range 0.04–0.06 per cent.

In addition to regulations on individual aircraft and on fuel, aviation is constrained by standards and objectives for local air quality (LAQ). For regulation, measures are based on concentrations, not on amounts emitted. This takes into account different meteorological conditions (which govern dispersion), background concentrations and building location, configuration and architecture, all of which will affect the exposure.

Objectives and observations are defined on varying time-scales. Often there is more than one definition for a specific pollutant; for example, a target may be set both for the peak value and for the annual mean. Using this range of time-scales allows the nature of health and environmental impacts to be reflected in the design of standards.

While some individual states have set separate objectives for air quality in the past, new EU legislation will become binding in 2010. The need to comply with this presents a significant constraint on airport expansion, including at Heathrow, where the need to mitigate air quality impacts has delayed consent for a third runway.

The dependence of emissions on engine operating characteristics means that reductions can be targeted through operational measures, but that not all measures will reduce all emissions. Some measures (outlined above) to reduce aircraft noise exposure on the ground can also have air quality benefits; for example, NPRs or increased angles of approach (or descent) could contribute to the achievement of air quality objectives, particularly by avoiding low-altitude flight over other sources, such as busy roads.

Improving the efficiency of taxi routing (perhaps through airport redesign) could also contribute to reduced emissions.

5.4   Climate Change

In recent decades, the evidence that human activities are significantly contributing to changes in the climate system has grown strong. These atmospheric changes, and their environmental and social consequences, will increase in the decades to come unless measures are taken to stabilise concentrations of greenhouse gases. Stabilisation would require emissions far below current levels, but exploiting technological and other opportunities to reduce our impact on the climate could reduce expected changes and slow the rate at which they occur. This section discusses how aviation, and particularly the management of air transport, contributes to anthropogenic climate change – and how it can contribute to the mitigation process.

The climate system is highly complex, driven by the interaction of physical and chemical processes. The emissions and impacts of aviation play a small but growing part in the human disruption of these processes. The trade-offs between different mitigations present serious challenges to reducing the net climate impact of aviation, particularly in the context of the current rates of growth in demand.

5.4.1   Carbon Dioxide (CO_2)

Carbon dioxide has been placed at the centre of the climate change debate; industrialisation was founded on the combustion of fossil fuels, releasing stores of carbon that had been sequestered over millennia. In the atmosphere, carbon dioxide acts to trap the heat energy radiated by the earth, warming the surface. This greenhouse effect has made the planet habitable, but human activity is now disrupting the process. Emissions of CO_2 (from sources including fossil fuels and land use change) have raised atmospheric concentrations from 280 ppm (parts per million) at the end of the eighteenth century to about 380 ppm.

The lifetime of carbon dioxide is long; much of the CO_2 emitted due to aviation over the past hundred years remains in the atmosphere. Atmospheric mixing means that concentrations are effectively uniform and the impact of 1 kg of CO_2 does not depend on the altitude or geographical location of emission, so emissions from aviation can be compared with those from other sources.

In 2004, air transport used 188 million tonnes (Mt) of fuel (Kim et al., 2005), emitting almost 600 Mt of CO_2 – 2 per cent of the 29 000 Mt emitted globally through use of fossil fuels (Marland et al., 2007). Locally, aviation can play a much larger role in total emissions. In the UK, for example, the share is 6.9 per cent (Defra, 2007). This share is expected to rise; aviation continues to grow while other industry sectors are constraining growth in their CO_2 emissions under the requirements of the Kyoto Protocol and other initiatives.

There have, though, been measures to reduce emissions from aviation. Engine fuel consumption for aircraft entering production at the close of the twentieth century was 40 per cent lower than that of the Comet 4, which entered production 40 years earlier. The transition from piston to jet aircraft carried a considerable fuel penalty, however. Only now are jet aircraft delivering fuel consumption per seat-km as low as that of the last piston-powered aircraft. In addition to the improvements in engine technology, increases in aircraft size and aerodynamic improvements have improved efficiency further; the fuel burn per seat for modern aircraft is around 30 per cent of that of the Comet 4. The load factor (proportion of seats occupied) has also increased, particularly with the rise of low cost airlines. This reduces the fuel burn (and hence CO_2 emitted) per passenger-km.

Despite these improvements in efficiency, growth in air traffic movements means that CO_2 emissions from aviation continue to increase. Figure 5.2 shows the CO_2 from international aviation for the six European countries reporting the highest values (United Kingdom, Germany, France, Italy, Netherlands and Spain) for 1990 to 2004. These six states together accounted for over 80 per cent of the total European Union emissions from international aviation in 2004. The source data for this figure are values reported to the United Nations Framework Convention on Climate Change (UNFCCC) and are based on the amount of fuel sold. The figures represent international departures only. To place European CO_2 emissions from international aviation into a wider context, it should be noted that emissions from the United States are highest, having reached 60 Tg, almost double that of the UK. Japan is the third highest emitter, with values reaching 20 Tg. International aviation emissions from China are not reported to UNFCCC. While China is ranked second in the world for total passenger-km when domestic operations are included, international operations are ranked fifteenth (ICAO, 2006) and the CO_2 from international aviation from China is likely to be similar to that of Italy or the Republic of Ireland.

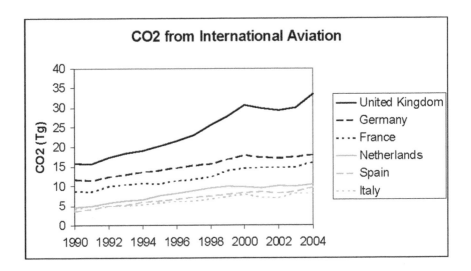

Figure 5.2 CO_2 from aviation for six highest emitting European countries

The rises in CO_2 from international aviation shown in Figure 5.2 should be seen in the context of changes in emissions from other sources. Table 5.2 shows the percentage change in CO_2 from international aviation and in the national totals for the same period (the national total values do not include international aviation). Values are shown for the six European countries with the highest CO_2 from international aviation. For all countries, growth in CO_2 from international aviation is much larger than the growth in CO_2 emissions from other sources.

It is also instructive to compare the carbon intensity of each mode – how much carbon is emitted per unit of productivity. Carbon emissions per passenger-km (or freight-tonne-km) are much higher for short-haul than long-haul air travel because of the high fuel requirements in the take-off and climb out phases, compared to

Table 5.2   **Changes in international CO_2 emissions**

Country	CO_2 % change from international aviation (1990–2004)	CO_2 % change national total emissions[a] (1990–2004)
United Kingdom	111	−5
Germany	53	−14
France	83	6
Netherlands	131	14
Spain	178	55
Italy	96	13
European Community	87	4

a   These totals exclude international aviation.

the relatively efficient cruise phase. For very long journeys, the fuel required per passenger- or tonne-km rises again; the additional fuel required to sustain long distance flight reduces the payload that can be carried. Comparisons between air and rail are clearly sensitive to the power source for rail, but, in general, values for medium-haul flights are larger than for the same distance travelled by high-speed train (powered by coal-fired electricity). For short journeys, where check-in procedures and the distance of airports from city centres means high-speed rail can often offer highly competitive journey times, the CO_2 from air approaches double that of rail. Comparing aviation emissions with road, for short-haul flights the carbon emitted has been estimated at roughly equivalent to each passenger making the same trip as a single occupant in a light truck (Penner et al., 1999).

Comparing carbon intensity per km is not the only approach. Passenger per trip values are much higher for air, as the distance travelled tends to be much greater. For the transport of freight, air transport emissions far exceed maritime and rail alternatives per tonne-km. The high cost of airfreight (partly related to its fuel intensity) means that its use is largely constrained to perishable or high value/ low weight goods. It is also worth noting in passing, that most freight booked as airfreight within Europe, is actually trucked by road.

The Kyoto Protocol was produced in 1997 and provides a framework for emissions reductions for six greenhouse gases, including CO_2. The agreement came into force in 2005. It has been ratified by 160 countries. Other countries, including the US and Australia, have not ratified and are not bound by the terms. The Protocol calls for controls on the emission of the specified gases in the commitment period 2008–2012, relative to emissions in 1990, with the reduction separately defined for each country. For the European Community, the agreed reduction is 8 per cent. Developing countries have no restriction on emissions under the Protocol. Through the Clean Development Mechanism, countries can choose to achieve their emissions targets by funding certain emission reduction or offset schemes overseas.

The terms of the Kyoto Protocol include CO_2 emissions from domestic aviation in national targets. International aviation is not included. The protocol called on developed countries to agree strategies for reducing or controlling emissions from international aviation through ICAO.

The most recent statement of ICAO policy relating to aviation emissions is set out in Assembly Resolution A35-5, which presents a consolidated statement on the organisation's position on a range of environmental issues. The resolution recommends more research and expresses a preference for technological measures over market-based solutions.

Table 5.3   Optimisation opportunities to improve fuel efficiency

Route	Aircraft	Practices
Optimise speed	Match type to route	Increase load factor
Optimise altitude	Improve maintenance	Reduce unnecessary fuel
Optimise taxi route		
Optimise flight route	Reduce empty aircraft mass	Reduce non-revenue flight

Source: ICAO, 2004.

Table 5.3 summarises the operational opportunities to minimise fuel use recommended in ICAO report AN/176 (ICAO 2004). Of these, the most significant fuel savings are expected to come from improvements to communication, navigation, surveillance and air traffic management (CNS/ATM) systems, which will improve the capability to fly the most fuel-efficient routes with optimised speeds and cruise altitudes.

There are considerable political and institutional barriers to policies to charge international aviation or to constrain its growth. Measures need to be global or multinational to avoid a simple switch in demand away from the country imposing them. In the absence of substantive policies from ICAO since the call for action of the Kyoto Protocol, policies have been explored at the level of the European Union.

Of the policies considered, inclusion of aviation in a future version of emissions trading has received most attention. Such schemes are designed to cap total emissions (in this case emissions of CO_2) at a maximum acceptable level and to allocate permits for emission. Participants able to reduce their emissions can then sell unwanted permits to organisations wishing to exceed their allocation. The approach ensures that emission reductions are achieved at the cheapest possible price.

The current EU emissions trading scheme began in 2005 and covers a range of industries including combustion plants, oil refineries, coke ovens, and the manufacture of iron, steel, cement, glass, lime, brick, ceramics, pulp and paper. Together, these account for about half of EU CO_2 emissions. The European Commission has announced plans to include aviation in the existing emissions trading scheme before 2012, but there are some policy issues yet to be fully addressed. One concern is how (and whether) the scheme should apply to airlines from outside the EU operating through EU airports.

In 1999, the Intergovernmental Panel on Climate Change (IPCC) produced a special report on aviation and the global atmosphere. This was the first such report focusing on one industry sector. Air transport receives special attention in the climate debate for a number of reasons. It is an international industry; emissions cannot be easily allocated to national quotas and effective regulation requires global cooperation. Those emissions also continue to grow rapidly, despite technological advances, due to aviation's important and increasing role in international trade and economic growth. Further issues arise because indicators of efficiency (such as emissions per passenger-km) can be ambiguous for mitigation.

5.4.2   *Nitrogen Oxides (NO_x)*

However, the key factor drawing aviation into the climate debate is that the non-CO_2 emissions and impacts of air travel are unique because of the altitude at which they take place.

Most emissions from aviation take place in the upper troposphere, close to the tropopause. The tropopause is defined by its cold temperature: in the troposphere, temperatures decrease as altitude increases. In the stratosphere, the warming from absorption of solar radiation by ozone means temperatures increase. The tropopause marks the boundary between these two regions. Emissions can change the radiative balance of the atmosphere by reflecting incoming solar radiation (which reduces the amount heating the surface) and/or by absorbing outgoing terrestrial radiation (which reduces the energy escaping to space). The key greenhouses gases (i.e. those trapping outgoing radiation) relevant to aviation are carbon dioxide, ozone, methane and water vapour. As discussed above, carbon dioxide emissions are a function only of the fuel burned, and impacts are independent of where that emission takes place. This is not the case for the other greenhouse gases produced by aviation. A further climate impact arises from the condensation of the exhaust gas to form contrails.

Nitrogen oxides are formed as a product of the combustion process and these increase ozone concentrations, thus contributing to warming. At cruise altitudes in the mid-latitude Northern Hemisphere, this increase is estimated at 6 per cent for 1992 relative to normal amounts, and is predicted to increase to 13 per cent by 2050. Impacts tend to be localised near major flight paths. The increase mitigates increased exposure to UV-B radiation through this reduction in ozone at higher altitudes. Between 1970 and 1992, the UV dose rate in northern latitudes has increased by 4 per cent; aviation has decreased this by 0.5 per cent.

Aircraft NO_x emissions decrease methane by increasing the rate of removal (due to increased OH concentrations). As methane is a greenhouse gas, this reduction leads to cooling. Methane is estimated to be 2 per cent lower than it would be in an atmosphere without aircraft, but this decrease occurs against a more than doubling of methane from other sources since pre-industrial times.

Nitrogen oxide emissions at altitude are not currently regulated, but legislation addressed at improving air quality close to airports has focused on reducing NO_x emissions during landing and take-off, which has also led to reductions in cruise emissions.

NO_x emissions do not necessarily reduce with improved fuel efficiency. Indeed, some fuel efficiency measures can increase NO_x emissions. One way to reduce specific fuel consumption (unit of fuel required per unit of thrust produced) is to increase the engine bypass ratio (the ratio of air passing through the fan ducts to the air passing through the engine core). Increasing the bypass ratio also reduces noise. However, one consequence of increasing engine bypass ratio is that NO_x production in the exhaust increases, so improved combustor technology is required to offset this.

5.4.3   Contrails

Contrails – the thin, white cloud tracks that sometimes form behind aircraft – cover about 0.1 per cent of the Earth's surface; in Europe, coverage is 0.5 per cent. Worldwide coverage is expected to grow to 0.5 per cent by 2050. The coverage from cirrus clouds formed from spreading contrails is not well understood and could be much larger – they begin as linear tracks, but can spread to form larger cirrus clouds (Figure 5.3).

Figure 5.3   Contrails begin as linear tracks

Contrails both contribute to global dimming (preventing solar radiation reaching the surface) and to global warming by acting as an insulating layer. The net impact varies according to many factors, including time of day (at night, the cooling effect due to reflecting sunlight is not present, so there is only a warming effect). Other influencing factors include the location and the microphysics of the contrail (the size/shape of the particles). On average, the net impact of both contrails and cirrus clouds for climate is understood to be warming.

Contrail formation depends upon the humidity and temperature of the surrounding air, and shows seasonal variability. A contrail forms when water droplets form on soot and sulphuric acid particles emitted from engines, then rapidly freeze due to the extremely cold temperatures at cruise altitudes. Contrails cannot form above a threshold atmospheric temperature, which reduces with altitude, as shown in Figure 5.4. The shaded area indicates the range of temperatures where contrail formation will depend on atmospheric humidity; the threshold maximum temperature is for saturated air. Typical atmospheric temperature profiles at low-, mid- and high-latitude (shown for March in Figure 5.4) fall below the threshold temperature at altitudes coinciding with typical aircraft cruise altitudes.

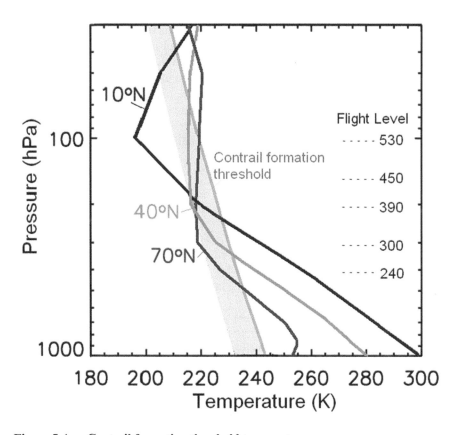

Figure 5.4   Contrail formation threshold temperature

There is strong evidence that contrails can form with activation of background particles already in the atmosphere (i.e. without soot or sulphuric acid particles in the exhaust). Ice particles in contrails take up water vapour from surrounding air (only a small fraction of the ice in a contrail-cirrus is formed from water from the engine exhaust).

When the atmosphere contains ice-supersaturated air masses, contrails can trigger the formation of extensive cirrus clouds. About 10–20 per cent of mid-latitudes are covered by these super-saturated air masses. Natural cirrus covers 30 per cent of the Earth at any one time. Aviation increases this by about 0.2 per cent, estimated to increase to 0.8 per cent by 2050.

Contrail (and cirrus) growth will tend to be faster than growth in CO_2 emissions, as most air traffic growth will occur in the upper troposphere, where conditions are more favourable for contrail formation. In addition, more efficient engines in the future will have lower exhaust temperatures for the same amount of water vapour emitted. This will allow contrails to form over a wider range of altitudes and at higher ambient temperatures.

Soot and sulphate aerosol have a small direct effect compared with other impacts. Both contribute to contrail formation, but may not be necessary for it.

A small fraction of the water vapour released by aircraft goes into the dry stratosphere, where it will accumulate and contribute to warming. Tropospheric emissions of water vapour do not have a significant impact, due to removal by precipitation. Supersonic aircraft emit water vapour directly into the stratosphere, where the lifetime is long and the radiative impact much larger, so any substantial transition to a supersonic fleet would change the nature and magnitude of the climate impacts of aviation. The increased warming would be partly offset because at these altitudes, NO_x emissions deplete rather than increase ozone. Despite this, radiative forcing (see next section) is estimated to be five times larger than equivalent subsonic aircraft.

5.4.4   Radiative Forcing

Radiative forcing is the measure usually used to compare the impacts of different mechanisms affecting climate. It is a measure of the amount of additional energy trapped in the troposphere. The measure is a global, annual average. It is used because it is approximately proportional to the change in global average surface temperature; positive radiative forcing indicates warming, negative indicates cooling.

While the measure is commonly used, it does not fully allow comparison of the different spatial and temporal patterns of impacts. This is particularly relevant for aviation, where impacts can be short-lived and localised (e.g. cirrus clouds) or long-lived and global (e.g. CO_2). Most aviation occurs in the Northern Hemisphere. Both CO_2 and CH_4 (methane) are long-lived and well-mixed throughout the atmosphere, so they impact both hemispheres. Ozone impacts and particularly contrail/cirrus have much shorter lifetimes, so impacts are focused along the main flight corridors of the mid-latitude Northern Hemisphere.

Table 5.4   Comparing mechanisms for aviation's impact on climate

Atmospheric effect	Lifetime	Radiative forcing (mWm^{-2})*	Understanding*
Carbon dioxide	Decades	25.3	Good
Ozone	Months	21.9	Fair
Methane[a]	Years	−10.4	Fair
Water vapour	Weeks[b]	2.0	Fair
Sulphate	Weeks	−3.5	Fair
Soot	Weeks	2.5	Fair
Contrails	Hours	20	Fair
Cirrus	Hours	30[c]	Poor

*   Radiative forcing for 2000, results from TRADEOFF (Sausen et al., 2005).
a   Methane is a greenhouse gas. Aviation emissions of NO_x speed up the removal of methane, reducing the concentration and thus leading to a cooling impact.
b   Most emissions occur in the troposphere where the lifetime is 1–2 weeks. In the stratosphere the lifetime is much longer and the warming impact much larger.
c   Estimated mean radiative forcing, with upper bound given as 80 mWm^{-2}. Considerable uncertainty remains.

Table 5.4 provides a comparison of the different mechanisms showing the radiative forcing and an order of magnitude for the lifetime and geographical scale of the impacts. Radiative forcing estimates are based on the calculations presented by IPCC for 1990. The confidence estimates indicated are those provided by IPCC. Note that the total, not including the effect of cirrus clouds (for which confidence is described as very poor), is more than double the impact of CO_2 alone.

Since the publication of the IPCC report, several studies have suggested that the impact of contrails may be lower than previously thought, that the impact of cirrus clouds may be larger, and that the combined contrail-cirrus impact may be the largest contributor to the radiative forcing of aviation.

5.4.5   *Climate Impact Mitigation: Achievements and Opportunities*

Aviation is a highly regulated industry, which contributes to its exceptional record of safety. The global networks of regulation have also driven environmental improvements, including reductions in emissions during the landing-take-off cycle. Nevertheless, several characteristics unique to aviation have presented challenges to including its impacts in international agreements on climate change. Firstly, spatial issues are significant. While both noise and surface air quality impacts are associated with the airport and its immediate environment, cruise impacts are harder to attribute. A further restriction on potential policies is that the network of bilateral agreements between states which allow the operation of international flights preclude the charging of tax on aircraft fuel. Debate continues on whether these bilateral agreements also preclude an environmental charge or levy to provide a further incentive for efficient fuel use. Fuel taxes or charges can be applied to domestic aviation, as demonstrated by the Netherlands and Norway.

High and rising fuel prices encourage improved fuel efficiency. The cost of fuel is not the only incentive, however. Reducing the weight of fuel required allows increased payloads or longer routes. These incentives have driven aircraft and engine design, so that each generation of aircraft uses less fuel per passenger-km or freight-tonne-km (with the exception of the transition from piston to jet aircraft, as mentioned, where the much lower price of jet kerosene compared to aviation gasoline allowed a considerable rise in fuel per passenger-km). In the medium term, this reduction in fuel per unit of productivity (averaged over the global air transport fleet) is expected to continue at the rate of around 1 per cent per year (ICAO, 2004).

One key challenge for the regulation of the climate impacts of aviation is that no incentives are in place to reduce other impacts. There is no direct benefit to the airline of reducing NO_x emissions at cruise, or preventing the formation of contrails.

In July 2005, European aviation industry groups including the Association of European Airlines (AEA) and the AeroSpace and Defence Industries Association of Europe (ASD) produced a joint position paper on emissions containment (AEA et al., 2005) arguing for technological and operational measures to be considered alongside economic policies. The document presents formal objections to economic instruments such as fuel taxes or emission charges, which it describes as 'blunt instruments with marginal environmental benefits but high negative economic impact' (AEA et al., 2005). It gives qualified support to emissions trading schemes, subject to demands including that aviation be included in a global and open system and that the scheme should focus only on CO_2 emissions. These conditions could potentially eliminate any environmental benefit from including aviation in such a scheme. By allowing open trading with other industries, aviation would be likely to become a net purchaser of emission permits. Allowing aviation to buy permits to offset its CO_2 emissions, either from another emitting industry making reductions or from a carbon sequestration industry such as forest plantation, while taking no account for the additional climate impacts from aviation from NO_x emissions and contrail-cirrus formation, would result in a net increase in the climate impact associated with each CO_2 permit.

Arguments have been made for the use of an 'uplift' factor to account for the additional impacts of aviation. This would allow 1 kg of CO_2 emitted by aviation to be defined as being equivalent to a larger amount from other (CO_2-only) sources. Values quoted for this 'uplift' factor vary widely. While this approach goes some way to address the additional climate impacts from aviation, it is still flawed. Firstly, the values are calculated from global mean annual radiative forcing. As we have noted, this is a measure of the additional energy trapped in the global climate system. As discussed above, the temporal and spatial patterns of the different impacts of aviation differ considerably and radiative forcing is only one measure of comparison. Different multiplication factors would be obtained if the future lifetime of the impacts was taken into account. Secondly, the level of confidence in estimates of radiative forcing by the different mechanisms also varies. Most current estimates exclude cirrus cloud impacts because of this uncertainty, so to describe the uplift factor as a reflection of total impacts is misleading. As a result of the weaknesses of this uplift factor, attempts to include non-CO_2 impacts of aviation in market-based measures to reduce emissions have been labelled premature (Forster et al., 2006).

In addition to market based measures, changes in air traffic management could also mitigate climate impacts. Many air traffic control sectors in Europe have boundaries aligned to national borders, with each nation taking responsibility for the airspace above it. The result is a complex routing network. Some measures have been taken to rationalise the situation, including the development of the Maastricht Upper Area Control Centre (Maastricht UAC), which manages high altitude air traffic over the Netherlands, Belgium, Luxembourg and part of Northern Germany.

Routes flown often differ significantly from the shortest (great circle) distance (GCD) between departure and arrival airports. Potentially, a route diversion could reduce emissions by optimising for the prevailing wind conditions, but in general, extra distance travelled means greater fuel consumption and more emissions. The opportunities to reduce fuel use through operational changes have been explored in a recent circular by ICAO (ICAO, 2004) – these were summarised in Table 5.3. While many of these opportunities relate to the business practices of the airline (including aircraft selection, load factor, fuelling practices and maintenance procedures), the benefits offered by communication, navigation, surveillance and air traffic management (CNS/ATM) measures are identified as the most significant. Improvements to CNS/ATM have the potential to allow more direct routing and to reduce airborne delays. (Potential improvements to aircraft flight paths and routing structures were discussed in Chapter 3 and will be explored as a key theme in Chapter 7).

There is further potential to reduce the climate impact of European aviation by eliminating or reducing contrails. Contrails can last for several hours and can spread to form large sheets of cirrus cloud. As mentioned, this plays a dual climate role, with a net warming impact. Studies to estimate the impact of contrails on a global scale continue, but recent results suggest that the combined impact of contrail and cirrus could be the largest contributor to radiative forcing by aviation.

In order for a contrail to form, the surrounding atmosphere must be sufficiently cold and moist for the expanding gas from the aircraft exhaust to rapidly condense and freeze. In general, reducing cruise altitudes across Europe could reduce contrail. There are several challenges to this. The first is the variability in atmospheric conditions. In winter, altitude restrictions would need to be much more severe than in summer (Williams et al., 2002), but the contrail reductions gained by applying benefits based on average monthly conditions would be threatened by the high day to day variability in conditions (Williams and Noland, 2005). Secondly, reducing cruise altitudes forces aircraft to operate at less efficient altitudes, increasing the fuel burn, which in turn increases both the carbon dioxide emitted and the cost to airlines. For longer haul routes, particularly, the journey time penalties could be significant, increasing airline staffing and other costs and raising problems for the predictability of schedules as the diversion required will depend on local variability in atmospheric conditions. The costs associated with delays and lack of predictability form the central theme of Chapter 4.

High resolution data have revealed that the supersaturated layers in which contrails can form are very thin (just 0.5km), raising the possibility that a monitoring system on board aircraft could identify contrail formation and advise an appropriate altitude change (Mannstein et al., 2005). As the altitude changes are smaller and more focused than those associated with any blanket altitude restriction applied across large areas, the penalties for fuel, journey time and airspace congestion would be significantly reduced.

Potentially, such a system could be extended to monitor background NO_x levels to include the impacts of emissions in the advice as to whether to change cruise altitude. Crucially, this would require international consensus on the prioritisation and comparison of the different mechanisms, which is not clear.

5.5   Trade-off Issues: Comparing and Prioritising Environmental Impacts

The prioritisation of impacts affects how policy decisions are, and should be, made. In the context of climate change, the time-scale used to assess cumulative effects determines which mechanisms are considered to have the largest impact. Prioritisation is also a problem when comparing climate impacts with other effects such as noise, particularly where measures to mitigate one impact can worsen another (Brooker, 2006). Attempts have been made to place an economic value on the various effects to allow comparison, but such efforts require subjective judgements and can be inequitable. The bulk of aviation emissions take place in and between affluent nations, while the consequences of climate change are likely to be worst for those developing nations least equipped for adaptation. This contrast inevitably leads to discussions of contraction and convergence, with the argument that policies should incorporate mechanisms to distribute resources more equally, rather than uniformly constraining rates of growth. The application of taxes or restrictions on aircraft type and equipage could particularly restrict new or existing key routes operating out of developing regions. The need for special consideration of airlines in these circumstances has been acknowledged by ICAO in their development of a balanced approach to noise management, but also applies to policy initiatives like the proposals to introduce emissions trading for flights operating to/from the European Union. Even within affluent nations, there are equity issues associated with the distribution of the benefits of access to aviation. Despite the recent rapid growth in low cost airlines, air travel remains largely the preserve of those on middle or high incomes (see Chapter 6 for a discussion of these issues).

The relative priority given to short-term local impacts, like noise and air quality, compared to long-term global impacts, like climate change, will influence judgement of the relative impact of long- and short-haul flights. This is important in shaping policy, particularly economic measures, as changes in the price difference between routes could influence traveller behaviour. These issues are also relevant to the development of hub and spoke business models versus direct routing (again, refer to Chapter 6 for further discussion on this).

These priorities cannot be objectively set and applied globally, or even across Europe; differing public perceptions change the identification of key environmental issues. For example, as noted above, the nuisance of noise emissions varies with affluence. Other factors, including the level of local and national development, will influence whether that nuisance is considered an acceptable consequence of increasing air travel (a theme explored in Chapter 8).

5.6   Looking to the Future: Impact Mitigation using Innovations in ATM

New technologies and concepts in air traffic management will offer opportunities for the mitigation of environmental impacts, either by reducing flight inefficiencies or by enabling optimisation to minimise environmental impact. New approaches to air traffic management can reduce flight inefficiencies by:

*   cutting dependence on ground-based navigation systems (so reducing deviations from the shortest (GCD) route)
*   reducing in-flight diversions from the planned route (for example for conflict avoidance or resolution)
*   allowing planned routes (such as Noise Preferential Routes, NPRs) to be followed more accurately

Air traffic management innovations could also contribute to environmental impact mitigation by improving the optimisation of flight trajectories to minimise emissions or impacts. These benefits could come from a wide range of areas in air traffic management research.

One area likely to offer significant opportunities for mitigation is communication, navigation and surveillance (CNS). With improvements in precision navigation (see Chapter 1), aircraft will be able to follow planned flight trajectories more closely. This will play a significant role in impact mitigation procedures, such as avoidance routes for approach and departure to minimise noise exposure. Other improvements in communication and surveillance could transform the relationship between pilots and controllers, allowing pilots to take more responsibility for selecting their own flight trajectories and maintaining their separation from other aircraft (see Chapter 3).

The ability to predict and avoid potential conflict situations at an earlier stage could also play a role in improving efficiency, by reducing (unwanted) diversions in flight. It would be a significant component of any scheme to optimise flight trajectories to minimise environmental impacts. Such measures would be likely to increase traffic congestion at some flight levels, particularly if contrail avoidance is included in the environmental optimisation. Medium-term conflict detection could also redistribute the workload on controllers, which is likely to become an increasingly constraining issue for airspace as traffic movements continue to increase (again, see Chapter 3 for further discussion on this).

Alternative approaches to conflict resolution, like lateral offset and speed control, could also change the environmental impact of flights. Lateral offset shifts an aircraft to a track parallel to its flight plan. This allows conflict resolution or avoidance using only small diversions from the initial route. Fast-time simulations have shown that, applied by a multi-sector planner, this technique could resolve the majority of conflicts associated with forecast traffic for the congested core European region for 2010 and 2025 (Ehrmanntraut, 2005). The effectiveness of manoeuvres like lateral offset would be further increased by improvements in navigation, as described above. Further fast-time simulations should be used to determine the relative fuel and emissions impacts of lateral offset compared to resolution manoeuvres currently in use.

Speed controls have the potential to resolve a large fraction of conflicts, but are rarely applied in en-route upper airspace in Europe, partly due to the long 'look-ahead' time required to determine when a speed control should be most effectively applied. A first assessment of their associated emissions has been made, and concluded that when traffic increases to double the base case (1997) scenario, using speed restrictions to strategically resolve conflicts could also reduce emissions, but further assessment of environmental impacts (particularly compared to other resolution methods) is required (Ehrmanntraut, 2005).

Speed controls could also be used to impose fuel efficiency measures on airlines. Improved understanding, by controllers, pilots and airline dispatchers, of the optimum operating speeds for aircraft in order to minimise fuel consumption could provide significant emissions reductions. This measure is among those identified by ICAO as potentially contributing to reductions in emissions (ICAO, 2004) (see Table 5.3). Optimising speeds for minimum fuel burn would result in some increase in journey times as aircraft are slowed to their most efficient operating speed. While the cost of fuel provides a strong market incentive for airlines to minimise their fuel costs, pressures to reduce journey time are also significant and result in less fuel-efficient flight operations (see also Chapter 4 for a discussion of this in terms of delay recovery).

Radical changes to airspace structure are also being explored. One innovative approach being researched is dedicated 'highways' for air traffic, which would contain several lanes of air traffic travelling along parallel routes defined to be clear from other traffic (Guibert and Guichard, 2005). The intention is to separate high-altitude, long-distance traffic from other traffic using continental-scale specified routes which link sections of airspace. These highways would be designed to reduce en-route delay on highly trafficked routes by having fixed access and departure points and procedures, and strict constraints. The exclusion of crossing traffic would reduce air traffic control workload and en-route delays and diversions associated with conflict resolution. Dynamically designing the highways for wind conditions would optimise performance based on fuel efficiency and/or journey time, but there is additional scope to include assessment of contrail formation conditions in the optimisation to further reduce impacts.

New visualisation techniques are required for controllers managing increasing numbers of aircraft. Current research is developing new approaches to human machine interaction for controllers (Bourgois et al., 2005), allowing dramatic changes to the conventional control screens and more manipulation of the field of view and the display, even by voice recognition. These tools may eventually be used for real-time air traffic control, but their introduction would require improved understanding of the human factors issues, including the ability of the controller to retain situational awareness in the event of system failure (as touched upon at the end of Chapter 1). As a planning aid, the visualisation opportunities offered include the overlaying of weather data, which could include contrail formation regions and/or data on NO_x emissions and impacts, to aid in an environmental optimisation process. This could also facilitate the design and selection of routes for highways.

Beyond air traffic management, innovations in aircraft technology and in airport location decisions will also contribute to efforts to control the environmental impacts of aviation. Some innovative research is being dedicated to the area of intermodality, to improve interactions between air and surface modes, and particularly between air and rail. This may have implications for the balance between short- and long-haul traffic, the environmental impacts of which depend on the metrics used for comparison (Williams and Noland, 2006). One longer-term aspect of intermodality that has been proposed is the opportunity to separate airside and landside operations, if security concerns can be addressed. This could potentially allow large hub airports to serve a number of cities. With these cities extending the catchment area for the hub airports, this would allow key routes to operate with high occupancy on large aircraft and so optimise efficiency. The relative environmental benefit of such schemes would depend on many factors and any assessment must include the wider environmental impacts of both modes, including noise exposure, land take and the energy source for rail, in addition to the potentially negative effect of non-rail access within such greatly increased airport catchment areas.[2]

Innovative approaches to route charging policy could also encourage environmental best practice, enabling external environmental costs to be better included in route planning.

5.7   ACARE – Setting the Research Agenda

Every two years, the Advisory Council for Aeronautics Research in Europe (ACARE) produces a strategic research agenda, setting objectives for European air transport in five key areas, including the environment. The high-level targets defined in 2004, for 2020, set out bold changes. The wider objective of an 'Ultra Green Air Transport System' is characterised by specific aims relating to different

---

2   Catchment areas as a function of market development and trends will be discussed in Chapter 6.

Table 5.5   Industry sector goals for 2020

Industry sector	Contributions by 2020
Airlines	New aircraft: quiet and lower emissions Preference for large aircraft and high load factors Avoiding extreme stage lengths Optimised speeds for fuel efficiency Introduction of Blended Wing Body aircraft Cooperation to reduce airborne delay Better environmental practice for servicing
Airports	Building efficiency and waste management Environmentally sensitive de-icing Hazard management to reduce fire risk Emissions limits on ground vehicles Low emission access transport Reduced taxiing and minimal use of jet engines on the ground Automated servicing Dedicated freight airports Precision take-off and landings Improved land use planning
Aircraft	Reduced fuel burn through aircraft, engine and systems design Reduced complexity laminar flow technology Aircraft configuration changes Adaptive structures New engine concepts for: noise, emissions and efficiency trade-offs New combustion and injection systems Availability of alternative fuels if climate benefit demonstrated Noise shielding Capability for steeper take-off/approach Reduced noise and emission rotorcrafts Lifecycle environmental impacts included in design process
Air traffic management	Green routes give incentive for new technology Routes dynamically optimised for atmospheric conditions Environment signature of aircraft registered in flight plans 4D trajectories for global traffic environmentally optimised Minimal diversion from agreed trajectories Real-time monitoring of environmental budgets

Source: ACARE, 2004.

actors in the air traffic system, as outlined in Table 5.5. These specific aims build on the ambitious targets from the first strategic research agenda,[3] which were:

- 50 per cent reduction in fuel consumption and CO_2 emissions;
- 50 per cent reduction in perceived external noise;
- 80 per cent reduction in NO_x;
- substantial progress in reducing the impact of aircraft manufacture, maintenance and disposal.

These targets refer to new aircraft to be introduced in 2020 and address the system as a whole, with contributions to come from a range of sources. Engine improvements are expected to give a 15–20 per cent improvement in fuel efficiency, with the remainder to be achieved through changes in the airframe and in ATM (Rolls Royce, 2003).

5.8   Conclusions

Aviation can bring considerable social and economic benefits, but imposes impacts on both the local and global environment. Significant progress has been made in reducing the impacts per aircraft near to airports, but as this has often allowed increases in traffic the total impacts have not been contained. For global impacts, the contribution of fuel to operating costs has encouraged innovation, which has reduced carbon dioxide emissions. At the same time, legislation to protect air quality near airports has had a secondary benefit by reducing NO_x emissions in the cruise phase. The achievements in reducing the environmental impacts of aviation have been considerable, and existing and proposed ATM concepts and tools offer opportunities for further improvement. However, there are two main challenges.

The first lies in determining the priorities for targeting impact reduction measures (both between climate change mechanisms and between environmental impacts) and ensuring that these are set on a sound scientific footing. Otherwise, measures could have potentially costly unintended consequences. For example, controlling only carbon dioxide emissions (either by technological means or by market measures like emissions trading) could increase the net climate impact if NO_x emissions and contrail formation rise as a result. There is also a danger that policies to discourage short-haul trips due to their higher carbon emissions per passenger-km could transfer some demand to long-haul, with its greater per trip emissions and impacts.

The second challenge lies in identifying the rates of industry growth that can be sustained and where that growth can and should be accommodated. Even to maintain existing levels of environmental impacts without any reduction, current growth forecasts suggest technological advances would need to be far greater than the ambitious goals set by ACARE. This is a complex societal question evoking

---

3   Advisory Council for Aeronautics Research in Europe (ACARE, 2002).

issues of equity within and between nations and generations – just how much growth in the environmental impacts of aviation are we prepared to accept for ourselves and for others?

These issues must be considered in context. Grounding all aircraft would not solve noise pollution, poor air quality or climate change, or the health and social problems that arise from those impacts. However, for each of these issues, and particularly in the context of climate change, it is increasingly apparent that while constraints are being faced by other industries, the case for allowing unconstrained growth in aviation will face growing environmental and political challenges.

Chapter 6

The Future of European Air Transport Operations

Nigel Dennis
University of Westminster

6.1   Introduction

Airline deregulation has had a dramatic impact on air transport operations in Europe over the last decade while at the same time, technological developments have altered the economics, size and range capabilities of commercial aircraft. Low-cost airlines represent the most notable new business model to have come into prominence in the last few years and these have proved amazingly resilient in the downturn following 11 September 2001 and through recent fuel price hikes and security scares, which saw many traditional airlines[1] losing large sums of money. For a general review of the market after this event, see Alderighi and Cento (2004).

Firstly, this chapter aims to address the special characteristics of the low-cost airlines with particular reference to changing route networks and the implications for infrastructure. The competitive strategies of the traditional airlines, which focus around frequency maximisation, hub strengthening and alliance development, are then reviewed. Finally, a look is taken at the opportunities presented by a new range of long-haul aircraft along with some perspectives on the growth potential of different market sectors.

6.2   The Growth of Low-Cost Airlines

Since deregulation of the US domestic air transport market in 1978 there has been much interest in how to create a scheduled airline that has significantly lower costs than its mainstream rivals and hence can viably offer much lower average fares. The main inspiration for this came from Southwest Airlines, which started way back prior to deregulation in 1971 as an intra-state carrier in Texas and has

---

1   Also referred to as 'legacy airlines' (especially in the US context), 'major airlines' or just 'majors'. The term 'flag carrier' is also often used to denote the largest airline of a country (historically often state-owned). None of these terms are precisely defined. We will use the term 'traditional airline' to differentiate from a 'low-cost' airline (which we shall thus avoid implying is not a 'major' enterprise!).

grown to become the largest US carrier in domestic passenger terms, transporting 1.5 million passengers every week – ahead of airlines such as American, United and Delta. Southwest has also been one of the most consistently profitable US operators, despite offering the lowest fares in the markets it serves.

It was inevitable, therefore, that when the deregulation focus turned to Europe, attempts would be made to create low-cost carriers in the European arena also (Lawton, 2002). This was accentuated by the starting position in Europe where the cost and fare levels of the traditional airlines on short-haul routes were among the highest in the world.

Prior to 1995, there were only a few Ryanair services at the low-cost end of the market in Europe. The low-cost airlines expanded to carry 33 million passengers in 2002, overtaking British Airways' short-haul traffic. Since then, a wide range of new entrants have joined the market – all claiming to be 'low-cost'.[2] Estimates put European low-cost traffic at 100 million passengers in 2004, and with growth rates of around 20 per cent per annum it may be more like 140 million passengers in 2006. This compares with 239 million short-haul passengers in 2004 on the Association of European Airlines' (AEA) member carriers (AEA, 2005), which includes the traditional airlines, but not all the regional carriers or the charters, which may account for another 100 million passengers between them.

A number of different business and product features have been adopted by these new entrants to reduce costs (Doganis, 2006). The number of seats in the aircraft is maximised in order to minimise the cost per seat. easyJet squeezes 149 seats into a B737-300[3] (or, more recently, B737-700) compared to 126 on British Airways. Comfort is not a key consideration on short-haul journeys. Further savings come from: increased aircraft utilisation; reducing the number of cabin crew to the safety minimum; no frequent flier credits; direct sales only (making extensive use of the internet, see also Chapter 4); and no transfers or interlining. In-flight services (e.g. newspapers, food, drink) are charged for, or eliminated altogether. Ryanair and flybe also now charge for checked baggage.

6.2.1   Fast Turnarounds

Low-cost airlines cut turnaround times by avoiding airbridges and using both front and rear stairs, with passengers walking across the apron to and from the gate. This makes aircraft types such as the MD80 series[4] or Fokker 100, with only a front entrance and a long thin cabin, unsuitable for low-cost operations. Having non-allocated seats speeds up boarding as passengers take the first seat

---

2   The definition of a 'low-cost' (or 'no-frills') airline has become rather blurred. Many, such as Air Berlin and bmibaby, really fall somewhere between the likes of Ryanair and a traditional airline like British Airways, while flybe lies somewhere between Ryanair and regional carriers such as Lufthansa City Line or Air Nostrum.

3   In naming conventions, 'B' stands for Boeing and 'A' for Airbus. In ICAO nomenclature, 'B737-300' becomes 'B733' etc. Also common, 'MD', is for McDonnell Douglas.

4   See footnote 3.

available rather than searching for, and blocking, others. Catering is minimised to avoid loading and unloading – this also saves on cleaning time (any rubbish is collected before the aircraft lands). Cargo is not handled, eliminating another source of complication and delay.

6.2.2   Secondary Airports

Table 6.1 shows how Ryanair ('FR'[5]) in particular has made use of low-cost secondary airports a critical part of its strategy (Barrett, 2000). Airport charges can be cut by negotiating favourable contracts with smaller airports that are desperate for traffic. Savings are also possible on station costs and ground handling, especially if an inferior level of service (e.g. long-queues at check-in) is accepted, although much of this work often has to be contracted out to a third-party handling agent which is likely to charge much the same to any carrier. Extra ground facilities such as business lounges can clearly be avoided.

Table 6.1   Examples of use of secondary airports by low-cost carriers

Major airport	Secondary airport
Rome Fiumicino	Rome Ciampino (U2, FR, NB)
Glasgow	Prestwick (FR,W6)
Paris CDG/Orly	Beauvais (FR)
Berlin Tegel	Berlin Schoenefeld (FR, U2)
Toulouse	Carcassone (FR)
Frankfurt	Hahn (FR)
Milan Malpensa/Linate	Bergamo (FR, LS, WW, AB, 8I)
Brussels	Charleroi (FR,W6)
Manchester	Liverpool (U2, FR, W6)

When buying a Ryanair ticket the passenger may well be deposited 70 miles short of the main destination but people appear willing to tolerate longer surface journeys in order to obtain lower fares.

Secondary airports improve on aircraft and crew utilisation by cutting down on queuing and congestion (see Table 6.2 – an example of the London-Frankfurt market, based on 2004 schedules: the last year a Gatwick service operated). Many low-cost carriers schedule virtually no contingency allowance (or 'buffer' – a key theme of Chapter 4) which means that delays can build up across the day. It is interesting to question whether easyJet has 'lost the plot' by expanding at Gatwick, one of the most congested airports in Europe!

---

5   Other table IATA codes: AB-Air Berlin, LS-Jet2, NB-Sterling, U2-easyJet, WW-bmibaby, W6-Wizz, 8I-MyAir.

Table 6.2   Secondary airports and fast turnarounds (2004)

Route	Block time	Turnaround	Output per 14-hour day
Stansted-Hahn	1 hr 15 min	25 min	8 sectors
London City-Frankfurt	1 hr 30 min	30 min	7 sectors
Heathrow-Frankfurt	1 hr 35 min	45 min	6 sectors
Gatwick-Frankfurt	1 hr 50 min	45 min	5 sectors

Low-cost carriers can make secondary airports work in a way that traditional airlines cannot by bringing in passengers from a much wider catchment area (the future environmental impacts of such changes having been discussed in Chapter 5). When Stansted had operations only by Air UK and some of the other traditional airlines, it effectively operated as a regional airport for people in East Anglia and, to some extent, East London. It was a bizarre choice for anyone from the much larger markets in central London or further west because it had lower frequencies than Heathrow, similar fares and a longer access journey. Now, with average fare levels half those of Heathrow, people are willing to drive past their nearest airport to fly from Stansted. This is particularly true of leisure passengers who appear to place a very low value on their own time.

Table 6.3 illustrates the example of Charleroi airport, where a survey of Ryanair passengers suggested that only 18 per cent of passengers resident at that end of the route came from the natural catchment area of Charleroi (i.e. southern Belgium). The rest were attracted from the rest of Belgium, the Netherlands, even Germany etc. If only a traditional airline service was on offer from Charleroi, these passengers would have used other airports such as Brussels, Luxembourg or Amsterdam. Without a low-cost service many of these people would not have travelled at all meaning that the market for a regional jet service from Charleroi may be as little as 10 per cent of the demand that Ryanair has captured.

Despite these handicaps, in certain markets, low-cost carriers and secondary airports are now handling a major share of traffic (see Table 6.4). For example, on London-Nice, easyJet has almost 50 per cent of the market (flying from Luton, Stansted and Gatwick). On London-Venice, Ryanair's services between Stansted, Luton and the secondary airport of Treviso have a 47 per cent market share and easyJet carries a portion of the Gatwick traffic. On London-Turin, Stansted and Luton both out-carry Gatwick. On London-Glasgow, although Heathrow still dominates, Stansted-Prestwick (a 'nowhere to nowhere' combination) has over half a million annual passengers and the low-cost airlines collectively have a majority of the local traffic (much of British Airways' demand is for transfers at Heathrow and Gatwick). Dramatic growth rates, albeit from a low base, have been achieved at the airports dominated by low-cost carriers.

Although the low-cost airlines can bring previously undreamed of traffic to the secondary airports, this may not be profitable growth for the airport operators (Graham and Dennis, 2007). These airports hitherto had little or no traffic, putting

Table 6.3   Ryanair passengers at Charleroi by residential location

Region	Proportion of traffic
Brussels Area	25%
Northern Belgium	19%
Southern Belgium	18%
Netherlands	17%
Luxembourg	8%
France	7%
Germany	6%

Source: University of Westminster survey of Ryanair passengers.

Table 6.4   Traffic splits by route for selected London short-haul markets

2005 Passengers (000s)			
London-Nice		London-Glasgow	
Heathrow-Nice	544	Heathrow-Glasgow	1427
Gatwick-Nice	299	Stansted-Prestwick	505
Luton-Nice	296	Luton-Glasgow	452
Stansted-Nice	122	Stansted-Glasgow	436
		Gatwick-Glasgow	372
London-Venice		London-Turin	
Gatwick-Venice	382		
Stansted-Treviso	319	Stansted-Turin	133
Luton-Treviso	87	Luton-Turin	78
Heathrow-Venice	81	Gatwick-Turin	60

Source: CAA Airport Statistics, Route Analysis.

the airline in a very strong position to negotiate cheap deals, as evidenced by Ryanair's arrangements with Charleroi (Aviation Strategy, 2001), parts of which were subsequently prohibited by the EU. Whereas a conventional airport may charge around €10 per passenger, and ground handling has to be contracted in addition, low-cost airlines expect to pay as little as €2 per passenger at secondary airports, including handling. Although it is logical to attract such 'marginal' extra traffic when existing infrastructure is under-used, the problem eventually can become that the 'cart starts pulling the horse' and the airport has to build new facilities. At Luton, a new terminal and railway station were built to cope with easyJet's expansion, but the airline launched a much publicised criticism of the increase in airport charges that followed (at a claimed £5.50 per passenger, still much below the published rate). Ryanair has undergone a similar dispute with Aer Rianta in Ireland and airports group TBI[6] blamed a fall in profits on the increased market share of low-cost airlines at their airports. Some airports (e.g. Marseille and Geneva) have developed a 'low-cost' terminal with differentiated facilities and service standards, potentially cutting the passenger fee in half – but as the landing charge will still be the same, this only amounts to about a one-third reduction overall. It is also impossible to stop full-service airlines moving to the new terminal, which could upset the overall economics of the airport.

6.2.3   Aircraft Size and Load Factors

The low-cost carriers are relatively efficient users of runway and airspace capacity.[7] A Ryanair B737-800 at 80 per cent load factor carries 151 passengers, almost double the traditional airlines which averaged 79 passengers per aircraft on European routes in 2004 (AEA, 2005). easyJet falls in-between, with an average of 123 passengers per flight (CAA, 2006b). The charter airlines have an even higher throughput however: 190 per aircraft for First Choice and 222 for ThomsonFly in 2005 (CAA, 2006b).

Table 6.5 shows that at many of the regional airports in Britain and Ireland the average number of passengers per aircraft movement has gone up dramatically since 1998, when services were primarily with small regional aircraft. The irony of this, however, is that runway capacity is not generally in short supply at these locations – unlike at the major hubs.

6.2.4   Impact on Traffic

However, as the provision of low-cost services increases, catchment areas can contract again. Examples of this are already being seen in the UK where, for example, East Midlands had the whole of the central England and Yorkshire market for low-cost travel to Barcelona to itself when bmibaby started flying the route in 2002. Three years on, alternative low-cost Barcelona services have started from Leeds/Bradford, Manchester and Birmingham while go (now easyJet) also launched a rival service from East Midlands to Barcelona. Is the market becoming too crowded? After Ryanair weighed in with a service to nearby Gerona, both easyJet and bmibaby called it a day.

There is some evidence of low-cost airlines starting to cannibalise their own traffic as network coverage grows. This may be because there is a finite pool of people willing to travel more and more as fares fall, hence these have to be progressively shared amongst a greater number of possible airports or destinations.

---

6   Now part of the Spanish corporation, Abertis.
7   Standardisation on the B737 and A320 families also helps to minimise vortex separations.

Table 6.5   Changes in average aircraft load at UK airports

Airport	Passengers per air transport movement in 1998	Passengers per air transport movement in 2003
Belfast City	39	62
Belfast International	74	106
Bristol	34	63
Cardiff	27	59
Cork	39	62
Dublin	79	97
Glasgow	60	75
Liverpool	51	93
Manchester	67	69
Newcastle	44	69
Nottingham	36	72
Prestwick	53	93
Shannon	37	74

Source: compiled from CAA statistics.

Examples include North-East Italy where a proliferation of airports now have low-cost services from London (see Table 6.6) and the South of France, where Ryanair started off with services to Carcassone and subsequently added Nimes, Montpellier, Perpignan and Biarritz.

Table 6.6   Low-cost traffic between London and North-East Italy

Airport	1999	2000	2001	2002	2003	2004
Bergamo	–	–	–	186	502	577
Brescia	–	85	189	189	121	177
Genoa	70	130	154	150	145	148
Milan Linate	–	148	230	124	143	166
Milan Malpensa	–	202	23	–	–	–
Turin	122	179	190	144	141	159
Total	192	744	786	793	1052	1227

Source: compiled from CAA statistics (passengers, 000s per annum).

It is difficult to draw conclusive findings on traffic generation as these markets may have been targeted by the new entrants because they were seen as having the right conditions to grow rapidly – regardless of any low-cost service.

The University of Westminster has attempted to make a crude estimate of the impact of low-cost airlines in the UK between 1998 and 2001 (it is not possible to update this analysis as there is no longer a substantial control sample of markets without low-cost service). Compared with AEA average growth rates, traditional airlines operating to/from the UK have still grown but appear to be 7 million passengers short of the expected figure. Charters are 1 million short, leaving a residual 5 million that can be assumed to have been generated by the low-cost airlines out of their growth of 13 million passengers. A more detailed study on a route-by-route basis suggests the growth has been most dramatic in smaller markets with strong tourist potential, where traditional air services were likely to be high cost and hence demand was artificially suppressed in the past.

The extent to which market stimulation can be continued by low-cost airlines in the longer term is more debatable. London is an almost unique case, with large traffic volumes to almost anywhere – it is a tourist destination as well as a generator of outbound traffic. Paris is the other European city to meet these criteria but there is no equivalent airport to Stansted in terms of the spare capacity that existed there a decade ago, coupled with lower costs. On the upside, however, several of the traditional airlines are unlikely to survive at their current scale of operations in the longer term, leaving openings for a more efficient provider. The UK CAA produced a report (Civil Aviation Authority, 2006a) suggesting that although there had been some switching between airports and destinations, the net growth attributable to low-cost airlines is much less than their total traffic.

The geography of mainland Europe means the private car (and, increasingly, high speed rail, actively targeting airlines, or, indeed, working in conjunction with them to relieve airport congestion on certain routes – see Chapters 5 and 8) is a more serious rival than other airlines in many cases.

There are relatively few dense international routes in mainland Europe – of the 20 busiest European air routes in 2000, 11 involved London and a further eight were domestic. Paris-Madrid was the only other international route in the top 20. A number of European cities only owe their current traffic volumes to the existence of a hub (e.g. Amsterdam or Frankfurt). The contestable market for point-to-point services is likely to be much smaller, even allowing for traffic generation through price cuts.

Leisure routes are therefore likely to dominate elsewhere and bmi's decision to hand its East Midlands-Brussels service back from bmibaby to bmi regional suggests that some markets cannot be stimulated significantly by lower fares. Further evidence that some of the recently launched low-cost routes cannot fill a B737 comes from Ryanair's axing of Ostend and Maastricht and bmibaby's attempt to sub-contract certain Cardiff routes to Air Wales. The rather mixed performance of low-cost subsidiaries of traditional airlines is studied by Graham and Vowles (2006), and the broader social implications of low-cost air travel are explored in Chapter 8.

6.3   **Route Operation – Frequencies and Regional Aircraft**

6.3.1   *Service Frequency*

In a competitive market place, frequency is a key determinant of an airline's market share among the lucrative business passengers. Figure 6.1 shows the famous 'S-curve' relationship between frequency and market share in business markets. An airline with less than half the frequencies in a two-carrier market will attract a disproportionately lower share of the traffic. A similar situation arises in multi-airline markets. Traditional airlines have attempted to compete against each other and the low-cost sector by reducing aircraft size and increasing frequency, as this offers the best chance of maximising market share of the most valuable, high-yield traffic.

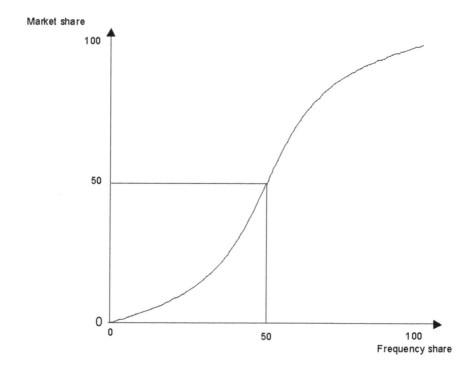

Figure 6.1   **S-curve of market share (two-carrier market)**

In long-haul markets, the key objective is to reach a daily frequency, although in a few very dense markets (such as London-New York) multiple daily frequencies are required to be competitive, which led to United's recent withdrawal as their one flight per day could not contend with BA's ten or Virgin's six. On many intercontinental routes, but particularly the North Atlantic, where there is limited freight demand, this has led to replacement of B747s at relatively low frequency with mid-sized aircraft such as the B767/B777, or the A330/A340.[8] This has also enabled more secondary gateways to be opened up.

On short-haul routes with a business focus, new entrants have tried to match frequency by using smaller aircraft types where necessary – which has driven all operators' aircraft sizes down and frequencies up. Table 6.7 shows the case of the London-Amsterdam route.[9] Although passenger traffic has tripled, frequencies have gone up by slightly more (from 19 to 61) and the average aircraft size has dropped from 134 to 115 seats, although some of this loss of seats is offset by higher load factors. Even at Heathrow, frequencies have more than doubled (from 11 to 25) but traffic has less than doubled – a questionable use of scarce slots.

Table 6.7   Developments in London-Amsterdam service

London airport	Monday–Friday, September 1984	Monday–Friday, September 2006
Heathrow	BA 4x757, 1xTRD	BA 8x320
	KL 4x310, 2xD9S	KL 8x737, 1x767
		BD 5x319, 3x320
Gatwick	BR 4xB11	BA 6x737
		U2 4x319
Stansted	UK 4xSH3	U2 3x319
Luton	–	U2 5x737
London City	–	VG 12xF50; KL 6xF50
Total frequencies	19	61
Average aircraft size	134 seats	115 seats
Annual passengers	1.1 million (1984)	3.3 million (2005)

Source: compiled from OAG and ABC data.

6.3.2   *Regional Aircraft*

Regional jets have proliferated in the United States in recent years as the traditional airlines seek to maintain frequency but reduce capacity in their main hub markets once local demand has been eroded by the low-cost carriers. These aircraft, such as the Canadair CRJ-200 and the Embraer 145, offer around 50 seats capacity and hence can tap markets that are too long for a turbo-prop and too 'thin' for a B737. The major snag is that the most viable new routes often involve a congested

---

8   See footnote 3.

9   Table codes: BA-British Airways, BD-bmi, BR-British Caledonian, KL-KLM, U2-easyJet, UK-Air UK, VG-VLM; B11-BAe1-11, D9S-Douglas DC9-30, F50-Fokker F50, TRD-BAe Trident, SH3-Shorts 330.

large hub airport at one end or the other. Although they initially offered seat-mile costs only about 20 per cent higher than a B737, high fuel prices are now making regional jets less attractive, particularly compared to turbo-props on shorter sectors. In Europe, there has been some 'tactical' use of regional jets, particularly by Lufthansa and Air France, while turbo-props have remained prominent in the fleets of KLM or flybe, for example.

Regional airlines still tend to have lower load factors on European services (50–60 per cent) than the traditional airlines (60–70 per cent); easyJet manages 80 per cent. This is mainly due to the peaked nature of the business traffic and whereas it is possible to attract German passengers to London for the weekend from Dusseldorf, for example, a route such as Manchester-Dusseldorf has very little generational potential. This is compounded by directional imbalances on many routes, such as Newcastle-Paris. It is therefore far from easy to make air services between two regional points a success and the average number of passengers per flight is very low at about 25–30 on a 50-seater aircraft.[10]

Regional aircraft have also enabled airlines to maintain a presence on routes not serving their major hub airport. On Manchester-Frankfurt, Lufthansa can easily fill four A320s a day with its volume of connections at Frankfurt. BA is in a much weaker position as this is effectively a point-to-point service for them. BA Connect, however, persevered on the route with Embraer 145s offering three flights per day (now a code-share, operated by flybe). Although Lufthansa will carry many more passengers in total, BA are able to mount a competitive service for the local market and, in particular, its many Frequent Flyer Programme (FFP) members in the Manchester area.

Short-haul business routes often have a lull in demand in the middle of the day. There are nevertheless some passengers who would benefit from a wider choice of timings. By utilising a mixture of aircraft types, supply and demand can be most closely matched. On Stuttgart-Paris, Lufthansa has a peak morning and evening flight with a B737 and a lunchtime and late evening service with an F50. This is possible by cross-scheduling the fleets: the B737 can perform a Mediterranean service midday and the F50 can operate a thin business route in the morning and evening.

A shift back from regional jets to turbo-props, especially the most efficient models such as the Dash 8 Q400 and the ATR72, is likely if fuel prices remain high. In Europe at least, turbo-props can be acceptable on much longer sectors than has hitherto been assumed, such as on Birmingham-Perpignan (704 miles; flybe). These are able to compete indirectly with low-cost airlines' B737s if using differentiated airports.

---

10   The UK CAA has produced a study of regional service developments (Civil Aviation Authority, 2005c).

6.4   Global Alliances and Network Restructuring

The strength of their network is one of the major competitive advantages for the traditional airlines over the low-cost new entrants (although this may have negative impacts on delay cost management – see Chapter 4). To take advantage of this an efficient hub-and-spoke operation is required. Hubs will therefore remain important, particularly for long-haul travel.

Alliances, franchising and code-sharing arrangements have become key means of consolidation in the international airline industry where bilateral agreements and foreign ownership rules prevent international mergers. In some circumstances, alliances can expand market coverage and increase competition, particularly where they are between two weaker carriers with complementary networks. In other circumstances, however, they may be more oriented to controlling capacity and reducing duplication or competition (Morrish and Hamilton, 2002). With many services around the world running at a loss they are one of the few means for the industry to address its (seat) over-capacity problems.

Airports and ATC providers will increasingly be negotiating with one or two powerful blocks of airlines. Table 6.8[11] shows how the Air France/KLM merger has effectively created three global alliances that dominate scheduled passenger traffic. The low-cost airlines are the other significant power block in the short-haul market.

At Heathrow, on an individual airline basis, half the slots are dispersed around a wide range of airlines, but when one groups them into alliance blocks, **one**world and the Star Alliance account for about 80 per cent of the airport's flights.

BAA has planned that Terminal 5 at Heathrow will be devoted to British Airways and Terminal 1 to the Star Alliance. At present, the (potential) alliance partners are scattered around the various terminals at the airport. The long lead times in airport infrastructure planning mean that new terminal facilities designed around one group of airlines may be obsolete by the time they are constructed.

At smaller hubs and secondary bases, the traditional airlines have become very dependent on franchises and regional partners to maintain network coverage. These have the advantage of lower cost levels and often smaller aircraft than the mainline operation. At Gatwick almost half British Airways' destinations are operated by a partner carrier, principally GB Airways on low-yield routes to the Mediterranean. Table 6.9 shows that at Lyon only six out of 42 Air France routes are by the mainline operation with the majority by Brit Air or Régional, using 50-seater aircraft.

Particularly for smaller airlines there is an attraction in combining with the appropriate foreign carrier in each market served. For example, Austrian's long-haul flights are all code-shared but with six different partners. This almost marks a return to the regulated era of pooling agreements when there was no effective competition on most international routes – instead the two flag carriers effectively operated a joint service.

---

11   Italicised airlines indicate the major partner in each region; Southwest and Emirates are the only non-aligned carriers in the world top 20.

Table 6.8   Major international alliance groupings (March 2007)

Star	oneworld	SkyTeam
United	*American*	*Delta*
US Airways		Northwest
Air Canada		Continental
Lufthansa	*British Airways*	*Air France*
SAS	Iberia	KLM
TAP	Malév	Alitalia
Austrian	Finnair	CSA Czech
bmi		Aeroflot
LOT Polish		
Spanair		
Swiss		
Singapore	*JAL*	*Korean*
All Nippon	Cathay Pacific	China Southern[a]
Thai	Qantas	
Air New Zealand	Royal Jordanian	
Asiana		
SAA		*Kenya Airways*[b]
VARIG	*LAN*	*Aeroméxico*
27% share of global RPK[c]	20% share of global RPK[c]	25% share of global RPK[c]

a   Joining.
b   Air France-KLM/Northwest partner.
c   Revenue Passenger-Kilometres.

Source: IATA, AEA, OAG and press reports.

Alliances can also spell problems for the level of air service at a particular airport (Dennis, 2005). Duplication, especially in Europe, is likely to be eliminated over time and although the key hub-to-hub trunk links are likely to see a rapid increase in operations, airports and routes which do not fit neatly into the alliance groupings are liable to see their service reduced.

6.4.1   The Long-Haul Context

Whereas most European countries can support a network of domestic and regional air services, long-haul traffic is much more concentrated. The four

Table 6.9   Franchises and code-shares: Air France at Lyon (2005)

Operator	Destinations
Air France	6
CCM[a]	2
Régional[a]	15
Brit Air[a]	16
Malév	1
Portugalia	1
Austrian	1
Total	42

a   Franchise/subsidiary.

Source: OAG and Air France website.

major hub airports (Paris CDG, London Heathrow, Frankfurt and Amsterdam) dominate the market.

Table 6.10 shows the overall picture in terms of all long-haul services.[12] Fifty-one airports in Europe had some form of long-haul service in July 2004. This ranges from one flight per week from Cardiff (to Toronto) and Hamburg (to Accra), up to 1125 flights out of London Heathrow (160 per day – some 22 per cent of the European total). Although Heathrow is well ahead in number of flights, Frankfurt and Amsterdam actually serve more destinations than Heathrow. This is primarily because there are more duplicated routes out of Heathrow – BA only has 40 per cent of the services there and, as well as foreign carriers, faces competition from Virgin in many cases. Heathrow also has some very dense routes such as New York JFK which accounts for 128 flights per week, or 18 per day.

Certain US destinations (e.g. Atlanta, Houston) are restricted under the bilateral agreement until March 2008 to operate only out of Gatwick and this accounts for another eleven destinations and 182 flights per week. This is because only airlines/routes already existing at Heathrow prior to the completion of the Bermuda II Agreement, almost 30 years ago, have been allowed to fly from there to the US. The route authorities of Pan Am and TWA were transferred to United and American in the early 1990s. With an 'open skies' (or EU multilateral) now provisionally agreed, these will almost certainly decamp to Heathrow, sending it to number one position in Europe for destinations and at the same time very greatly reducing Gatwick's remaining long-haul services, already depleted from

---

12   Based on the AEA definition which includes, from Europe, all Atlantic services, sub-Saharan Africa, Asia and Australasia. It does not include North Africa or the Middle East, which are classified as medium-haul.

Table 6.10   Long-haul services by European airport (July 2004)

Airport	Non-stop destinations	% Routes operated by hub airline or code-share partners	Weekly frequencies
London Heathrow	71	40	1125
Paris CDG	78	62	806
Frankfurt	81	69	671
Amsterdam	60	67	480
Madrid	30	54	276
London Gatwick	32	21	244
Rome Fiumicino	34	44	165
Zurich	25	70	164
Milan Malpensa	35	75	153
Munich	33	76	136
Manchester	18	19	108
Paris Orly	11	41	107
Others (39)			790
Total			5225

Source: Dennis (2007).

their heyday, by British Airways in 2000. We shall return to a discussion of the long-haul market later, but turn next to the specific context of the hub.

6.4.2   Hubs Remain Important

The major hubs have strengthened their position in recent years as previously significant competitors such as SAS and Alitalia have lost ground, while Swissair and Sabena have disappeared – to be replaced with much smaller-scale operations: Swiss and Brussels Airlines. BA has transferred Gatwick flights to Heathrow and Air France now has a minimal long-haul presence at Orly.

Overseas airlines tend to avoid the medium-sized airports that are only important as hubs (unless part of the same alliance group) e.g. Zurich, Milan, Munich, Vienna. Compared to ten years ago, some concentration is apparent. Thin, low frequency routes from regional airports have been dropped to boost the flows through the hubs. The largest markets have generally shown the most growth (Sweetman, 2004; p. 30). Whereas once cities such as Toulouse, Bordeaux, Lyon and Basel/Mulhouse had direct flights to New York, only a Nice link survives among the French regional airports, everything else being forced through Paris, or alternative hubs. Direct services such as Hamburg-Atlanta and Birmingham-Chicago have also disappeared.

Analysis conducted in the US suggests that hubs remain effective for many types of air travel demand (the differential performance between the hub airlines and the low-cost airlines has been exaggerated by restrictive labour practices – a large hub airline with a new entrant's terms and conditions would also be a formidable competitor – as some of the 'legacy'[13] carriers have demonstrated since reorganising under Chapter 11 bankruptcy protection.[14]

Hub airlines should be more efficient on routes from their hub airports where they uniquely obtain the economies of density in terms of connecting passengers. Thin markets, unable to support a two- or three-times daily B737 service on the basis of origin and destination traffic (even after price stimulus), are more efficiently served via hubs. Long-haul services remain the preserve of the traditional airlines for reasons that will be explored further below. Conversely, the point-to-point model works best on dense routes away from the major hub airports (consider many of the London services from Luton and Stansted) or holiday markets (the traditional charter business and, increasingly, the low-cost airlines' focus also).

Hubbing is advantageous to airlines because it increases the number of markets in which they can compete for a given volume of output. The network coverage grows exponentially as the size of the hub increases. This strategy has become more attractive to airlines with the advent of greater pricing freedom through deregulation. An airline such as KLM can compete in the Frankfurt-New York market by selling a service combining two of its own flights via Amsterdam.

This also offers important passenger benefits and potentially higher yielding traffic by providing access to regional airports (e.g. Humberside is linked with Amsterdam by KLM City Hopper providing the opportunity to make journeys such as Humberside-Milan or Humberside-Tokyo).

For airport operators, hubbing provides the only real opportunity to grow beyond the traffic potential of their local catchment area. This raises a number of issues, particularly concerning: capacity and its utilisation; the type of passengers served; the composition of the airline customers and the type of facilities that are provided. The non-hub airports will be affected also in terms of the services they enjoy and the knock-on impact of hub operating procedures.

6.4.3   Scheduling at Hubs

Scheduling is probably the most critical factor in operating an effective hub. There is little point in having superb airport facilities in a major city at the heart of Europe if passengers have to wait four hours for their connecting flight. In this time they could have flown another 3000 km.

If the passenger is prepared to wait an indefinite time at the hub, connections can be achieved between all services operating to and from it. In reality, long waits at the transfer airport are unattractive, especially where the actual flying time is short. If alternative routes are available, a considerable drain of traffic

---

13   See footnote 1.
14   For example United, US Airways or Delta – as analysed by Nuutinen (2007).

may be experienced. Even in a monopoly position, optional demand will still be suppressed.

Superficially, it appears to be the frequency with which individual routes are operated that will determine the waiting time at the interchange point. If a service operates every four hours, the passenger will, on average, have to wait two hours to catch it, plus any time required for being processed through the airport – the Minimum Connect Time (MCT). The transfer time becomes particularly critical in the context of air transport as a result of the low frequencies in relation to journey times and the long passenger handling times that are the rule. On routes to quite significant European cities such as Lisbon, Lyon or Stuttgart, British Airways only operates three flights per day from Heathrow.

An essential element of any serious attempt to maximise the scope of an airport as a hub is to concentrate activity into a limited number of peaks, or 'waves', during the day. These should see a large number of inbound flights arriving during a short space of time, then departing again as soon as a sufficient interval in which to redistribute passengers and their luggage has elapsed (Figure 6.2). The transfer time between flights in the same wave will be close to the best attainable. A further benefit associated with this type of scheduling is the ability to ensure connections are available in both directions in the main city pair markets as the timetable will be symmetric. This is not possible under random scheduling and will be a major consideration for transfer passengers who largely wish to make round-trip journeys.

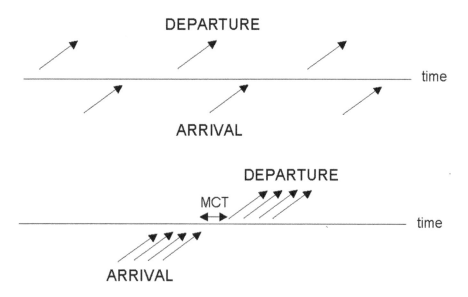

Figure 6.2   The wave concept

Inevitably, it is only the local airline and certain agreeable partners that will conform to this pattern. Operators not based at the hub airport have less to gain from the multiplier effects and will be more strongly motivated by requirements of the point-to-point traffic or their own hub systems elsewhere.

Ample runway capacity is an essential pre-requisite of a successful hubbing operation. Through waves of flights, severe peaking of arrivals and departures is necessary to optimise the availability of connections and some spare capacity is preferable to allow a margin for absorbing delays. Two contrasting examples will be used of different levels of hubbing intensity in Europe: London Heathrow and Amsterdam.

London Heathrow is an example of an airport where waves of flights are not operated. Instead, one runway is used for arrivals and one for departures. As the airport is running close to capacity this leads to an almost uniform pattern of activity across the day, with about 20 arrivals and 20 departures in each half-hour period. British Airways, the dominant airline, has around 40 per cent of movements in each interval. Terminal 5 (T5) will improve BA's connectivity by reducing the MCT but it will do nothing to improve the schedule or increase runway capacity. Indeed, T5 is not large enough to accommodate all BA and **one**world operations, so there will still be interchange required with Terminal 3.

At Amsterdam, KLM operates a connecting hub with three main waves of flights per day and an emerging fourth one in the mid-afternoon. Short-haul aircraft are mainly stabled abroad overnight, flying into Schiphol in the early morning. The other airlines using Amsterdam are more random in their pattern of movements although there is some further concentration around the KLM peaks, mainly from independent commuter airlines providing additional feeds, but also some long-haul carriers looking to take advantage of the European connection opportunities. The densest activity involves 31 KLM departures and nine others between 1900–1929. At least two parallel runways are required in-service permitting simultaneous arrivals or simultaneous departures, plus one contra-flow runway.

Whereas Heathrow has a steady pattern of movements all day, at Amsterdam there is around a three-fold increase in the peak period compared to the average. As well as requiring use of multiple runways, this poses major problems for airport operators in relation to terminal capacity. It also imposes a cost penalty on the hub airline because it obtains less than efficient utilisation of its ground staff and facilities. Everything from the number of terminal gates, baggage handling and check-in counters, to the number of air traffic controllers, must be designed to cope with the peaks.

It is also worth noting that to maximise connections it is necessary to stable aircraft away from the hub overnight. This is a major divergence from traditional European airline operating practices where almost all aircraft and crew would be based at the principal city and return there overnight. This can lead to a loss of maintenance and support services at the hub airport in favour of the spoke locations. These disadvantages can be more than offset by the revenue and traffic benefits of operating a hub.

6.4.4   The 'Continuous' Hub

One term which has recently entered circulation and is being vaunted as a new concept that will enable the traditional airlines to redress the cost imbalance with the 'no-frills' operators, is that of the 'continuous' hub. This effectively means abandoning the waves to give a uniform pattern of activity across the day. Connections will be created randomly by the sheer weight of flight activity at a busy hub. This is not a new concept – it is the old-fashioned way of doing things before the operational research specialists became involved.

The argument is that by keeping the aircraft moving all day rather than constraining them into fixed waves, utilisation and productivity can be increased. It also reduces staff and infrastructure requirements at the hub to cope with the artificial peaks that the waves create. It is further claimed that punctuality will improve, offsetting the disadvantage of longer waiting times at the hub. American Airlines has now started to roll out this strategy across its major domestic hubs after many years of favouring the concentrated wave structure. Others such as Continental, Northwest and most of the traditional European carriers have adhered closely to the wave system, however. The key question seems to be whether an improved quality of service can still generate extra traffic, or a market share or yield premium, or whether everything nowadays is driven solely by price – regardless of convenience and journey time.

At a busy hub with 450 departures per day grouped into approximately eight waves, the typical wave schedule, as traditionally operated in the US, would see 60 aircraft arrive in the space of 30 minutes (e.g. 0830–0900), all stay on the ground for at least 30 minutes and depart again within a further 30 minutes (0930–1000). Thus, the worst case arrival still offers 60 onward connections within 90 minutes waiting time. By directionalising the waves, so that aircraft move say east-west (or *vice versa*), this ensures that all these connections are potentially marketable. The process is then repeated at two-hourly intervals throughout the day.

Compare this with the 'continuous' hub. The same number of daily flights gives one departure every two minutes. From an arrival at 0830, passengers can then connect to 30 flights between 0900 and 1000 (15 of these between 0900 and 0930 will indeed be faster than in the above scenario). This immediately halves the number of connections available within 90 minutes, however. If the passenger is prepared to wait a further hour (150 minutes in total) then 60 connections become available again. However, whereas with the wave scenario these were directionalised to optimise their marketability, this is not possible with a random hub. It is likely that only half will be in the required direction, with the remainder being back-tracks. Thus the number of marketable connections falls to around 25–50 per cent of the wave hub (depending on the criteria one adopts for tolerated waiting times and direction). This is the reason why almost every traditional US airline has favoured the wave hub for the last 20 years. Its mathematical and geographic advantages are indisputable. Although it increases costs, this has always been by much less than the potential increase in revenue (e.g. a 10 per cent cost increase leading to a 30 per cent revenue increase). It is instructive to note how Air France's performance and profitability soared when they switched from a continuous hub to a coordinated hub with waves, in 1996.

It is also even debatable whether a continuous hub improves reliability for connecting passengers. It will relieve the over-peaking which leads to aircraft queuing for take-off but the contingency allowance that comes from the longer layovers ('at-gate buffers' were discussed in Chapter 4) required to maintain the wave cycle will be lost. With a high utilisation model, more flights are likely to be delayed, not fewer. Whereas the traditional hub is designed to ensure that a service in the inbound wave connects to a specific target group of departures, and action can be taken to assist a specific aircraft and passengers that are arriving late, or, indeed, to move the whole wave later when services are widely disrupted (again, see Chapter 4 for further discussion of this theme), there is no such guarantee with a continuous hub. Any possible connection is just one pairing out of many, and tight connections are unlikely to be maintained if the arrival at the hub is late. The onus is hence on the passenger (as with Southwest or easyJet) to allow more than the basic MCT if they want to be sure of catching their connection rather than being in a wave of flights that are designed to connect, putting the responsibility on the airline to make it work.

So when might a continuous hub be a good idea? For low-cost airlines it makes sense because with a large enough price differential (say 50 per cent) then sufficient passengers will be willing to put up with lengthy transfers at the hub (also highly necessary if there is no through checking of baggage). Southwest, for example, carries one third of its traffic as transfers or through passengers in this way.

At hubs which generate a large amount of origin/destination traffic, so that it is possible to fill most of the plane with local passengers, then 60 feeder flights in a wave schedule may be excessive and it is possible to still obtain sufficient transfer business to 'top-up' with a continuous hub. There are only a few cities in the world where this might be the case: London is one (and Heathrow manages around 30 per cent transfers, which seems to be the natural level for a random hub) and New York another (where Continental operates a fairly uniform/random hub at Newark that has been quite successful). However, even at Chicago, half the traffic is transfers and at other US hubs it is as high as 75 per cent. This level of transfer business could only be accomplished with a wave-type schedule.

The other situation where waves become unnecessary is where most routes operate at high frequency (eight plus, per day). Then a connection will always be possible within two hours, allowing for a 30-minute MCT. If the high fuel price environment persists, however, we may see a move back to lower frequencies with larger aircraft under an efficient wave arrangement to maintain connections.

6.4.5   Hubbing of Low-Cost Services

One of the characteristics of low-cost services is their lack of interline facilities with the traditional airlines. However, the attitude of the low-cost carriers towards intra-line connections (within each airline's own services) differs greatly.

Most of the European low-cost carriers actively discourage anything other than point-to-point traffic by selling sector fares only, with connections (e.g.

from Scotland or Ireland, via Luton or Stansted, to continental Europe) at the passenger's own risk. Only Air Berlin and Virgin Express have gone wholeheartedly for the hub-and-spoke concept.

Contrary to popular perception, most of the US low-cost carriers do sell connecting journeys rather than purely point-to-point. In the case of Southwest Airlines, there is no particular scheduling assistance but it is possible to link up between Southwest services at a number of major nodes (e.g. Dallas Love Field, Baltimore, Phoenix, etc) and attractive through-fares help to offset poor connection times. AirTran operates something akin to a conventional hub at Atlanta, as does Air Berlin in locations such as Palma (Table 6.11) and Stansted. The wheel is hence reinvented!

Table 6.11   Air Berlin hub at Palma

Origin	Palma arr.	Palma dep.	Destination
Frankfurt	0910	1015	Porto
Leipzig/Halle	0910	1020	Lisbon
Salzburg	0910	1020	Jerez
Dusseldorf	0915	1025	Bilbao
Dortmund	0920	1025	Malaga
Vienna	0920	1030	Barcelona
Munich	0925	1030	Madrid
Amsterdam	0930	1035	Alicante
Berlin Tegel	0935	1040	Seville
Hamburg	0935	1040	Valencia

Source: OAG (February 2006).

6.5   A Further Look at the Long-Haul Market

6.5.1   New Long-Haul Aircraft

Aircraft size has been drifting downwards for some years in the long-haul arena – most US carriers, for example, do not operate any B747s today, favouring B767s and B777s, especially on the North Atlantic and Latin markets. The latest move, however, is the use of significantly smaller aircraft than the B767 on very thin, medium-distance routes. The B757 can be adapted for transatlantic operations and is used on a handful of services, mainly by Continental. It also appears on short routes from Europe to Africa, such as Madrid-Lagos. The B757 offers the opportunity to return to an aircraft of B707 size (around 150 seats) but has the downside of only a single aisle with three seats either side.

Currently, there is some interest in whether a niche can be found with a small narrowbody aircraft such as the A319 or Boeing Business Jet – a long-range derivative of the B737. Lufthansa converted non-hub services from Dusseldorf to Chicago and Newark, and have since added Munich to Newark. Lufthansa/ United used to operate larger aircraft (A340s and B767s) on the Dusseldorf-US runs. The rationale is to retain the high-yield business traffic which is willing to pay a premium (i.e. the full business class fare) for a non-stop service. Although unit costs are high, so are unit revenues. Other passengers are forced through a hub (e.g. Dusseldorf-Frankfurt-Chicago).

The new wave of all-premium start-up airlines (Eos, MAXjet and Silverjet) have gone a step further and launched a discounted first or business class product from secondary London airports using full-size, long-haul aircraft such as the B767 (Nuutinen, 2005). These have shown greater staying power than some commentators expected. Although this concept may succeed on London- New York it is difficult to extend it to other city pairs. London-New York is such a large market, with many top-end passengers, that only a very small market share is needed and there are already many potential customers starting or finishing their journeys close to the secondary airports.

bmi British Midland have expressed an interest in operating long-haul services from the British regions where an A330 is too big. The bmi plan, using A319-LR or B737-700X equipment envisages a conventional two-class cabin, as there is insufficient premium traffic on routes from places such as Manchester.

For the future, Airbus and Boeing have taken a rather different prognosis of the requirements of the market. Both can currently expect to capture a significant part of the mainstream demand with their A330/A340 series and B777, respectively. Airbus believes that factors such as growth in demand, downwards pressure on costs, slot shortages at key airports and an increased dependence on hubs and alliances will push airlines towards larger aircraft: hence the development of the A380. Boeing, in contrast, believes passengers will want more non-stop flights on thinner and long-range markets with a cost effective, smaller aircraft: hence the development of the B787. Boeing sees the main market gap in the next ten years as a B767 replacement (ideally with a longer range) and is anxious not to lose customers to the mid-market Airbus models. Airbus has now proposed the A350, which represents an improvement on the A330 as a twin-jet rival to the B787.

It is debatable whether there is unmet demand for non-stop services that, until recently, were beyond the range of suitable aircraft. Most of these markets are between Asia and North America. From Europe, it is only Chile (from Northern Europe) and Australasia that fall into this category. London-Perth non-stop would be possible with the A340-500, but perhaps only viable if a sufficient volume of high-yield business passengers offset the higher costs.

The A380 will have some presence in Europe, mainly in Europe-Asia markets. Dense traffic, coupled with slot constraints, means that most of the Middle East and Asian operators intend to fly the new aircraft to London and some to other European cities also. In Europe-Asia there is already a sizeable B747 presence. Time-window constraints mean that some of these sectors have multiple B747s or A340s leaving on almost identical schedules (e.g. Hong Kong-London) – these could be efficiently combined onto a single aircraft. The 43 A380s on order by Emirates must imply hub development on a massive scale at Dubai. This should be a source of concern to every other airline operating between Europe and the Far East!

It is unlikely that many A380s will appear on the North Atlantic, although hub-to-hub routes within the Air France and Lufthansa alliances could be candidates. Virgin Atlantic may also use it on London-New York and there could be some fifth freedom services (these involve intermediate traffic on routes to/from the home country), for example by Singapore Airlines operating Singapore-London-New York. Heathrow is likely to be one of the major airports on the A380 networks, even though British Airways has not so far ordered the aircraft.

6.5.2   Scope for Low-Cost Long-Haul Airlines

The low-cost airline revolution has so far been entirely confined to the short-haul market. In the US, Southwest and Jet Blue fly some transcontinental routes but these are only medium-haul by international standards (five hours). There are a number of reasons why it is more difficult to translate the low-cost formula to the long-haul market (Francis et al., 2007), although it has been tried, most notably by Freddie Laker's 'Skytrain', on the North Atlantic, some 25 years ago!

Traditional airlines in general already obtain low seat mile costs and hence offer competitive fares on long-haul services. From London to New York, a typical winter advance purchase, economy class return, is as little as £200 including taxes, rising to £500 in peak season. Whereas in Europe, low-cost airlines have been able to more than halve the average fare paid per passenger, the best they are likely to achieve in long-haul is about 20 per cent off. In the long-haul markets there remains a significant market willing to pay a premium price for sleeper seats, etc. With passengers at the front of the cabin paying many thousands of pounds for their ticket, the marginal cost of the economy class seats at the back of a mixed configuration aircraft falls considerably. By filling the aircraft with economy class it would be difficult to do better than this, especially as seat pitch on long-haul cannot realistically be reduced below the 31″ or 32″ already provided by the traditional airlines. On some aircraft types it is possible to squeeze an extra seat across the cabin (e.g. eight abreast instead of seven on the B767, ten instead of nine on the MD11).

It is difficult for low-cost airlines to match the utilisation improvements that have been achieved on short-haul routes, as long-haul aircraft are already flying fifteen-sixteen hours a day with carriers such as BA and Lufthansa, many sectors being overnight. It is also difficult to eliminate 'frills' altogether. Some form of meal service is required on flights of eight or ten hours. Even if paid for 'on demand', the costs of the galley space and the complications of loading catering remain. Non-allocated seats are a no-go: families are unwilling to be split up for that length of journey. In-flight entertainment is also more important on long-haul than short-haul and the number of toilets realistically cannot be reduced from the traditional airlines' provision (as has been done on short-haul routes). Furthermore, large amounts of checked baggage must still be handled. Civair was a South African domestic airline intending to fly Cape Town-Stansted from the end of October 2004. Economy return fares were advertised from £420 not including food, drink or headsets. This was about the same price as indirect flights on Lufthansa or KLM and £150 less than the direct operators from Heathrow. The service never started.

Hubs are much more crucial for long-haul travel than for short-haul. The only dense, long-haul point-to-point markets from Europe equate roughly to Virgin Atlantic's network from London plus a handful of Paris routes and a few New York services. Other services are heavily dependent on connecting traffic at one or both ends of the route.

Use of larger aircraft than the traditional airlines would be necessary to reduce unit costs. Thus if BA is using a B777 it would be possible to undercut them on seat-mile costs with a new A380. This, however, flies in the face of the low-cost airlines' strategy on short-haul routes where they have kept to the modest B737-size equipment in order to remain competitive on frequency. Without hub feed, large aircraft are not really a viable proposition. Also, on long-haul, cargo is too significant a source of revenue to ignore, particularly if flying aircraft with large belly-hold capacity.

If these commercial obstacles were not sufficient, the regulatory barriers in the form of bilateral agreements limit the markets in which a new-entrant, low-cost airline could start service. UK airports (except Heathrow and Gatwick) have relatively liberal access to transatlantic routes and some Far East markets and the UK government would probably be supportive. In France, however, there is likely to be more protectionism of Air France. Several schemes have been mooted for linking Stansted with a US low-cost base such as Baltimore, enabling passengers to create their own 'low-cost' connections. It is difficult to see this being a very efficient process, however, with three – presumably independent – airlines involved.

A new entrant could undoubtedly find staff willing to work for less, although the differential is muted compared to short-haul routes. Traditional airlines often pay staff the same across the network, which makes them particularly uncompetitive on short-haul. On long-haul, low-cost airlines would still have to incur some overseas accommodation and allowances as it is physically impossible for staff to return to base each trip.

For these reasons, there are few long-haul charter flights, which provides some evidence of the constraints in the market. The only cases where charters have been successful with long-haul are on leisure dominated routes in the peak season. This is reflected by the low frequency, scheduled services operated by leisure airlines such as LTU, Martinair and Air Transat.

Although the circumstances are clearly loaded against a successful invasion of long-haul routes by new-entrant or 'low-cost' airlines, it cannot be ignored. With long-haul services remaining strongly profitable again for the traditional airlines then it is likely that other airlines will wish to get a slice of this market. If European traffic for the low-cost airlines falters, then it is possible that carriers such as easyJet may have to look at picking up interline traffic to supplement their own local demand or even operating long-haul in their own right.

6.6   Growth Trends and Forecasts

There are a wide variety of influences that generate a demand for air travel. In the business market these include economic development, international trade and the globalisation of industry and services. In the leisure market there are social factors, as people change jobs more frequently and work further away from home, creating a need to visit friends and family. Migrant workers from new EU member states such as Poland have stimulated air traffic to Britain in a similar way to the UK-Ireland market in the 1990s. An increase in income levels coupled with the falling real-terms cost of flying has made air travel more affordable (see also Chapter 8).

Exchange rates can work either way – when the US dollar is strong, North Atlantic travel is usually most buoyant, as many Americans come to Europe; when the dollar is weak there is not generally a balancing increase in the number of Europeans going to the US. Interest rates and inflation rates affect the balance between spending and saving. Leisure time is increasingly a constraint, especially in Japan and the US where holiday entitlement of only two weeks per year is not unusual. This reduces the scope for additional air trips outside the main holiday. Growth in international trade stimulates business travel in particular, while the political climate can make or destroy a destination.

We saw some EUROCONTROL traffic forecast scenarios earlier, in Table 3.1 (Chapter 3), and the three most influential factors to be used in a simple forecasting model are indicated in Table 6.12. Economic growth alone shows excellent historical correlation with growth in air transport. Fare levels and level of service (essentially the number of routes and frequencies, which provides a measure of the convenience and accessibility of air transport: not the in-flight service!) are also significant. The elasticity indicates the change in air travel demand that is expected per 1 per cent change in the named (explanatory) variable. For example, a 10 per cent cut in fares should raise demand by between 5 and 10 per cent, so the elasticity is –0.5 to –1.

Table 6.12   Key inputs to forecasting models

Explanatory variable	Elasticity
Economic growth (GDP)	+1 to +2
Fare levels	–0.5 to –1
Level of service	+0.1 to +0.5

Figure 6.3 shows the type of forecasts that result. Expected sustained growth in GDP (Gross Domestic Product) drives a strong increase in air travel. Most forecasters produce similar projections (the manufacturers are generally a little more optimistic than governments and airlines, for obvious reasons). However, one could observe that a flattening 'S'-type curve would also fit the data! Figure 6.3 is a world aggregate forecast; Table 6.13 sub-divides this by region. Boeing's

regions are slightly different from other organisations', such as IATA. North America and Europe are forecast to have lower growth rates but still keep ahead of the Asian markets by 2023. The Chinese internal market grows rapidly, as do other short-haul markets in North-East Asia and South America.

A number of significant markets are just on the cusp of income levels that can support mass air travel (e.g. India, China, Poland) and some more traditional markets (such as Italy) would still appear to have strong potential for growth to bring them up to their peer group average. Even within the UK, residents of the South East make more than twice as many international air trips as those of Wales. There may be something of a 'chicken and egg' effect here in relation to proximity to airports, as the North West (which includes Manchester Airport) is the second most travelled region. Obviously, income levels and cultural differences play a part also, and these may become more uniform over time (calculations from CAA airport survey, 2001).

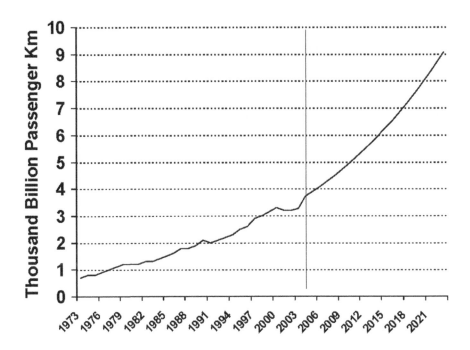

Figure 6.3   World airline passenger traffic forecast to 2024

Source: adapted from Boeing's *Current Market Outlook* (Boeing, 2006).

As touched upon earlier, whilst air is likely to remain the main mode of travel in Europe for journeys over about 800 km (500 miles), high speed rail services can make major inroads into the 500 to 1000 km segment, with private car dominating the shorter distances. Table 6.14 (which includes both UK and foreign residents) shows the air market share of all travel between the UK and

Table 6.13   Differential growth rates in major markets

Rank in 2005	Rank in 2025	Regional flow	% Growth in passenger-km p.a.
1	1	North America – North America	3.6
2	2	Europe – Europe	3.4
3	3	Europe – North America	4.5
4	4	China – China	8.8
5	5	North America – North-East Asia	5.8
6	6	Africa – Europe	5.0
7	7	Europe – South-East Asia	4.8
9	8	South East Asia – South East Asia	5.8
8	9	Central America – North America	4.1
10	10	North-East Asia – North-East Asia	5.4
19	11	South America – South America	6.9
14	12	Europe – South America	5.8

Source: Boeing (forecast average growth rates 2005-2025).

Table 6.14   Modal split of travellers between the UK and EU countries (2003)

UK to/from	% by air	% change since 1993
Belgium	25	–12
France	30	0
Luxembourg	58	–
Ireland	61	+8
Netherlands	63	+8
Germany	74	+18
Austria	87	–
Italy	90	+11
Denmark	95	–
Sweden	95	–
Spain	96	+4
Portugal	97	–
Finland	98	+3
Greece	99	–

Source: compiled from International Passenger Survey Data (published in *Travel Trends*).

various European countries. For the more distant locations, air has a share of some 80–90 per cent. Italy is slightly out of line (on the low side), whilst almost everyone flies to Spain.

About 15 per cent of total UK air traffic is domestic and this is the segment most vulnerable to surface transport competition in the future. More onerous security procedures, increasing fuel prices or taxes, improved rail services or motorway upgrades and more intercontinental flights from regional airports could all reduce the potential market for domestic air services. If the competitive balance does not change drastically, however, then there should be ongoing growth in the base market which will drive a continued growth in demand for domestic air travel.

6.7   Conclusions

Although ultimately it is the airlines that may determine the appropriate commercial provision of air services, this interacts to a considerable extent with the airport facilities and airport and airspace capacity provided. Indeed, it was remarked upon in Chapter 1, that whilst much has been said about the need for investment in airports to prevent a capacity crisis, if that spending is not forthcoming, this calls into question the need for high investment in ATC, as capacity will become increasingly constrained by runway and terminal availability.

The network of air services provided is altering as a result of structural changes in the industry, as well as competitive pressures and the development of passenger demand. The low-cost carriers are relatively efficient users of runway capacity as they maximise the number of seats in the aircraft and operate at high load factors. Their concentration on B737 or A320 series equipment is efficient in utilising airspace capacity as it minimises vortex separations. Some conflicts and complications to airspace routings, however, arise from the large number of regional and secondary airports that now have movements crossing the main air traffic flows, whereas most operations used to be focused on a small number of major airports. It is still the case, however, that the densest European routes are either domestic sectors or links with London. There are few large international markets in mainland Europe.

The low-cost carrier business model favours the use of secondary airports as they are less congested, enabling faster turnarounds and minimising queuing or holding. This can be offset, however, with a more demanding schedule incorporating less buffer time. Pressure is also being placed on the traditional airlines to increase aircraft utilisation at the expense of schedule reliability.

The traditional airlines are moving in the opposite direction regarding aircraft size at all but the most congested airports. This is because frequency is a key competitive factor for the high-yield business passengers. Furthermore, in order to maintain hub feed once the local traffic has been eroded by the low-cost carriers, smaller gauge aircraft are often required. In the long-haul markets too, increasing competition as a result of newly deregulated air services agreements is likely to lead to smaller aircraft in order to maintain or increase frequency. Boeing's 787 is designed for a fragmenting marketplace – although most of the new services will still be from a hub at one end or the other. Some niche products are also emerging using long-range A319s, or similar equipment. The counteracting factors are where airport slots are at a premium, when it is sensible to use as large an aircraft as possible, or if there is a large volume of cargo demand, since larger aircraft have relatively greater cargo capacities. On some routes, a lack of suitable time windows also eliminates frequency competition. For these services the A380 may be the more appropriate option.

Hubs present specific operational issues including the concentration of a large number of arriving or departing flights within a short space of time. This requires multiple parallel runways operated in mixed-mode configuration to optimise the throughput. Infrastructure and ATC facilities must be designed to handle the peaks rather than a more uniform flow. That is one reason why an airport such as Schiphol, which handles a similar number of aircraft movements to Heathrow, requires five runways rather than two.

It is superficially attractive to increase frequency and reduce aircraft size to spread the hub peaks while maintaining connectivity. If the high fuel price environment persists, however, we may see a move back to lower frequencies with larger aircraft under an efficient wave arrangement to maintain connections.

In a competitive market it is not easy to make the most efficient use of capacity. Competition has undoubtedly brought passenger benefits in terms of greater accessibility and choice but often undermines the infrastructure planning goals that could be achieved in a more regulated business environment.

Chapter 7

The Single European Sky – EU Reform of ATM

Ben Van Houtte
European Commission

7.1   Introduction

The Single European Sky is the European Commission's initiative for the reform of air traffic management. It started in late 1999 with extensive preparation for a regulatory package that was eventually adopted by the Council of Ministers and European Parliament in early 2004. From that time on, the Commission concentrated on elaborating detailed implementation rules to give effect to the principles and working methods set out in the legislation. A technological complement to the Single European Sky package was launched in 2006 under the 'SESAR' label. This chapter will set out the main elements of the EU's approach in this area and will provide some perspective on the priorities for the next few years.

The Single European Sky initiative touched on some politically and legally sensitive issues, such as the relation between a new Community policy and a continuing intergovernmental cooperation, the scope for EU legislation on the organisation of member states' (aerial) territories, and the relationship between civil and military organisation of air traffic. It came close to the limits of what can be achieved under the European Community's transport policy, and in that sense is a good case study of the challenges that present themselves for the EU in its further development.

7.2   The Organisation and Evolution of Air Traffic Management

As mentioned in Chapter 1, the legal organisation of air traffic management, globally, derives from the 1944 *Convention on International Civil Aviation* (also known as the 'Chicago Convention'; see Chapter 1). It gives states the right to decide over which portions of their territory[1] they wish to provide air navigation

---

1   The authority of states is normally limited to their land areas and territorial waters. However, on the basis of regional agreements approved by ICAO, states may be given the authority to provide air traffic services over high seas.

services. The Convention and its annexes set out some basic requirements with which these services have to comply, and the International Civil Aviation Organisation (ICAO) defines common standards and practices that provide the basis for an air traffic management system that is relatively homogeneous throughout the world – even though it sometimes tends to be a 'lowest common denominator', rather than reflect the state of the art.

While, traditionally, air navigation services were provided by government departments, almost all EU member states have now set up corporate entities for that purpose. Most of these are publicly owned, but some have been partly privatised or are being prepared for privatisation (with the consequent societal implications of changing ownership discussed in Chapter 8). At least for core air traffic services, these Air Navigation Services Providers (ANSPs) enjoy a natural monopoly. The area of responsibility for the ANSPs typically coincides with the territory of its state and any high seas that may have been entrusted to that state. In a number of cases, neighbouring states have agreed on boundary adjustments as a result of delegations of responsibility to provide air navigation services, but these remain limited in scope (refer back to Chapter 1 for further details, and see also discussion on Cross Border Areas (CBAs), below).

National ANSPs are not always a very practical approach, as the territory of states may be too small to allow for efficient and cost-effective ATM, and as traffic flows may not be accommodated easily in this territorial organisation. Therefore, ICAO has, for a long time, promoted joint air navigation service provision going beyond boundary adjustments and extending over the territories of several states. In a few cases, this cross-border approach was pursued in Europe. For example, Luxemburg has delegated the responsibility for most of its lower airspace to Belgium; and the upper airspace over the Benelux countries and over Northern Germany is managed through a joint facility set up by EUROCONTROL in Maastricht Upper Area Control Centre (Maastricht UAC). Nevertheless, the overall picture remains one of high fragmentation (as touched upon in Chapters 3 and 4). By way of comparison – where it must be recognized that the analogy is not complete – in Europe, 58 en-route centres are managing a similar geographic area, and less than half the traffic volume, managed by the US Federal Aviation Administration, with only 21 centres. No wonder that the unit cost of the European system is between 60 and 70 per cent higher than in the US (EUROCONTROL, 2003a). In addition, it must be acknowledged that the monopoly arrangements prevailing both in Europe and elsewhere are not conducive to cost reduction; the general rule is that ANSPs are entitled to recover their costs from airspace users through an elaborate charging mechanism.

As introduced in Chapter 1, EUROCONTROL, with a view to improving the coordination of air navigation services in Europe (currently between 37 states – including all EU member states, except for Estonia and Latvia) is working towards a seamless pan-European ATM system. The organisation, for example, initiated a number of programmes to introduce improvements to ATM systems across Europe (e.g. reduction of the vertical separation between aircraft – RVSM; see Chapter 1), and it provides several functions of common interest (e.g. it collects overflight fees and manages the flow of air traffic by delaying some departures to avoid overload of the system elsewhere – an underpinning principle of Chapter 2). EUROCONTROL, as an intergovernmental organisation, is not part of the EU family but is based on its own treaty.[2]

Since, in particular, following the adoption of the Single Sky legislation, most EUROCONTROL activities relate to subjects for which the EU member states have agreed to pursue a common policy, the European Community has acceded to EUROCONTROL in order to be able to exercise member states' obligations in these areas of Community competence.[3] As a result, EU member states are bound to act jointly on these issues. In practice, most of the technical work is done by member state and industry technical experts, but with coordinated decision -making with EUROCONTROL bodies. The European Commission normally speaks on behalf of EU member states on issues of Community competence, following coordination of the positions within the EU.

7.2.1   The Current Military Dimension

As we saw in Chapters 1 and 2, airspace is used not only by commercial and private flights (GAT) but also for military training and operations (OAT). Since it would be dangerous to mix general and operational air traffic, the military typically use segregated volumes of airspace for their training and operations and restrict access to these. We also mentioned that since the dissolution of the Warsaw Pact in 1991 the situation in Europe has changed considerably, with a general decrease in military activity, but with an increase in the *volatility* thereof.

Historically, due to the presence of air bases (pre- and post-World War II), many of today's military areas are still located in the core of Europe, within national boundaries, often competing for airspace with major commercial air traffic flows. As we saw in Chapter 2, the military may only activate certain types of segregated airspace on a temporary basis for the time they need to conduct exercises, but even then the disruption of air traffic can be severe, limiting the possibility of operating more direct civil routes.

The emergence of NATO has promoted the creation of partnerships with similar, or equal, weapons systems, and led to increased exchange programmes between air forces. Greater space requirements created by new weapons systems (and higher performance aircraft), together with the need for spaces for international air force training, have shifted the problem relating to the occupation of blocks of space in Europe from a national, to multinational, level. Whereas previously, bilateral training would take place inside a national boundary (e.g. the

------

2   The European Organisation for the Safety of Air Navigation was set up in 1960 and is currently governed by the Protocol consolidating the EUROCONTROL International Convention for the Safety of Air Navigation of 27 June 1997. The Protocol has not yet entered into force pending the ratification process, but is being applied on a provisional basis.

3   Council Decision of 29 April 2004 on the conclusion by the European Community of the Protocol on the accession of the European Community to the European Organisation for the Safety of Air Navigation, OJ 2004, L 304/209.

Swiss Air Force would train with the Norwegian Air Force in northern Norway), so-called Cross Border Areas (CBAs) have now emerged. These areas of military segregated airspace overlap national borders, and Letters of Agreement between the military authorities and civil ATC fix their terms of utilisation.

Several nations with large land and sea areas have shifted the training areas of their air forces towards spaces that are less used by civil aircraft in order to contribute to civil decongestion, partly caused by the concentration of civil traffic on some very saturated civil airspace cross-points. On the other hand, the introduction of the higher performance aircraft has also modified the nature of military requirements, such that they now have a greater need for high altitude, large areas, where much of the commercial traffic still takes place.

There is considerable scope for improving these arrangements, in particular by reducing the number of small areas of limited use to the military and by moving increasingly to larger areas, preferably those where commercial air traffic is less dense. However, the military clearly has a demonstrable and legitimate need for airspace, and it would not be right to give them only residual areas that may be far removed from air bases. Furthermore the location of military airspace cannot be modified easily because of operational, financial and social considerations. Therefore, the tendency so far has been to concentrate on improving arrangements for opening the same volumes of airspace to both civil and military air traffic at different times – through the Flexible Use of Airspace (FUA), as discussed in Chapter 2. Longer term, additional improvements could be pursued by further exploiting the scope for joint military operations, including CBAs, thereby reducing the need for each member state to have its own (often suboptimal) segregated areas. In addition, the location and number of airbases could also be modified, thereby providing opportunities for relocating the segregated airspaces.

Furthermore, the military often plays a major role in the conduct of GAT because they operate some of the facilities (e.g. mixed civil/military airports) and because they provide some air navigation services to civil airspace users (e.g. air traffic control in some areas, provision of meteorological information). The increased emphasis on security of air traffic following the events of 11 September 2001, underscores the need to develop close working arrangements between civil and military organisations so as to ensure the rapid detection of, and reaction to, threats that may originate from GAT.

For these reasons, it is necessary to develop a comprehensive approach including both the civil and military components of ATM. A few EU states (Germany being an example) have been able to integrate both organisations but, usually, efforts made by states and by EUROCONTROL towards that objective have produced limited results (EUROCONTROL, 2001a).

We will explore the military context specifically within the Single European Sky context later.

7.2.2   Community Involvement in ATM

The Community's air transport policy developed considerably since the 1980s to address substantially all elements relevant to the aviation industry (Van Houtte

2000; Ayral, 2003), starting with commercial issues relating to airline competition and air services, but gradually also extending to more technical areas such as safety and equipment. However, ATM remained mostly outside the ambit of Community legislation. ATM was raised occasionally by EU institutions[4] but the thrust of the Community's action was to let EUROCONTROL deal with these issues. Member states were no doubt generally satisfied with the action taken within the EUROCONTROL framework, perhaps also because they could not envisage ways in which to handle the military dimension of ATM within the EU context. As a result, the only Community legal instrument dealing with ATM was a directive making mandatory certain equipment standards developed by EUROCONTROL.[5] The EUROCONTROL convention was revised in 1997, in particular to make it possible for the Community to join the organisation and to exercise Community powers in the field of the free movement of goods, as well as to give EUROCONTROL some regulatory power.

In 1999, this approach started to change. Air traffic management suffered unusual disruption as a result of the military operations over former Yugoslavia (the 'Kosovo Crisis') and of operational changes in the route network. Delays reached critical levels (see Chapter 4) and the organisation of ATM was severely criticised in the press and at a political level. EUROCONTROL was not perceived to react effectively to this crisis (as will be discussed in Chapter 8): the organisation was blamed for not being sufficiently sensitive to the expectations of airspace users, for slow decision-making and for its inability, once changes were agreed, to put these into practice. At the same time, the Commission made the development of a Community initiative in the field of air traffic management, to complement traditional EUROCONTROL actions, into a priority, on the basis that because of its political mandate, effective decision-making and enforcement mechanisms, the Community was in a position to carry out a real reform of ATM.

In late 1999, the Commission issued a Communication on 'the creation of the single European sky'[6] which emphasised the need for structural reforms, with a view to integrating the management of airspace and developing new ATM concepts and procedures. The Commission announced the setting up of a high-level group, composed of senior representatives of civil and military air traffic authorities from the member states.[7] This approach was endorsed first by transport

---

4   Resolution of the Council and the Ministers for transport, meeting within the Council of 18 July 1989 on air traffic system capacity problems, in effect rejecting the Commission's proposals for legal instruments to deal with air navigation services, equipment and airspace set out in the Commission's communication (COM(88) 577).

5   Council Directive No 93/65 of 19 July 1993, OJ 1993, L 187, amended by Commission Directive No 97/15 of 25 March 1997, OJ 1997, L 95 and implemented by Commission regulations No 2082/2000 of 6 September 2000, OJ 2000, L 254, and No 980/2002, OJ 2002, L 150.

6   Communication from the Commission to the Council and the European Parliament: The creation of the single European sky, COM(1999) 614 final.

7   The Commission also invited Switzerland and Norway to participate in the discussions on account of the close involvement of these third countries in the Community's aviation policy on the basis of air transport agreements.

ministers during their Council meeting on 9–10 December, 1999, followed by heads of state and government during the Lisbon and Feira European Councils in March and June, 2000, and, finally, by the European Parliament in July, 2000.[8]

The high-level group met throughout the year 2000 and completed its report in November (European Commission, 2001b). In parallel, the Commission had convened an 'industry and social group', composed of representatives of the main stakeholder communities (airspace users, ANSPs, airports, trade unions and professional staff organisations, equipment manufacturers) in order to provide input into the discussions of the high-level group. The report of the latter emphasised the need to consider airspace as a common resource, managed as a continuum. Gradual progress was required towards joint civil and military management of air traffic. The group argued for more efforts to be made to develop and deploy new technology and improve the interoperability of equipment. It confirmed the desirability of involving professional and trade union organisations in the Community social dialogue. The reform of ATM required the intervention of a strong Community regulator, as the EU institutional framework was deemed to be the only suitable way to make rapid progress towards more efficient and coherent ATM. The group expected the Community's involvement to rely on synergy with EUROCONTROL's expertise in the area. In the course of the discussions, the Commission made it clear that its objective was neither to liberalise ATM (which, for technical reasons, continues to be a natural monopoly) nor to launch a privatisation process (for which the Community in any case has no authority). Its purpose was rather to develop a regulatory framework in order to enable effective decision making and implementation, so as to intensify efforts towards integration and modernisation of the system, and to deal with an industry which is increasingly operating as a business rather than as government departments and which pursues opportunities outside the home state.

7.2.3   Community Legislation

While the high-level group had been preparing its report, the Commission had launched a number of studies to analyse technical aspects of airspace and service provision. It maintained frequent contacts with stakeholders and endeavoured to build a constructive relationship with trade unions, *inter alia*, by launching a sectoral social dialogue on air traffic management[9] (see also Chapter 8). As a result, the Commission was ready to follow up on the high-level group report, and devoted most of 2001 to preparing legislative proposals.

---

8   European Parliament resolution of 6 July 2000 on the creation of the Single European Sky, OJ 2001, C 121/470.

9   On the basis of Commission decision No 1998/500 of 20 May 1998. The social dialogue aims to develop the social dimension of the Single Sky process, promote EU-wide agreements on working conditions. It deals with issues such as air traffic controller and other ATM personnel licences, change management in the ATM industry and reorganisation of air navigation service provision into Functional Airspace Blocks.

There was some concern, however, that the Commission's upcoming proposals would be held hostage to the long-running dispute between the United Kingdom and Spain on the application of Community legislation to Gibraltar Airport because of differences over the sovereignty of the territory in which the airport was located. At the time, this dispute held up a large number of air transport proposals in the Council. Eventually, a provisional understanding was reached between the authorities concerned, according to which the application of the Single Sky legislation to Gibraltar Airport would be suspended until a more permanent solution was developed[10] (a definitive agreement between the UK, Spanish and Gibraltar authorities on, *inter alia*, the use of Gibraltar airport was concluded on 16 September 2006[11]). On this basis, at the end of 2001, the Commission came forward with proposals for a framework regulation setting out the principles and working methods to be followed in the Single Sky, and for specific regulations dealing with air navigation services, airspace and equipment.[12] Discussions took about two years and were concluded at a conciliation between European Parliament and Council in December, 2003; the texts entered into force on 20 April, 2004.

7.2.4   Working Methods

The Single Sky legislation takes the form of regulations, obviating the need for extensive transposition measures by member states.[13] Nonetheless, national authorities continue to play a major role, not only because some of them still participate in the business of providing air navigation services, but mostly because the trend towards increasing corporatisation, or even privatisation, requires the development of adequate monitoring structures.[14] Therefore, member states are expected to set up independent supervisory authorities that operate at arms length from the ANSP. Where a state still provides these services itself, it has to ensure at least 'functional separation', meaning that safeguards are required to avoid supervisory decisions being affected by conflicts of interest, due to operational functions being carried out within the same organisation.[15] It should be noted that where several states participate in the formation of so-called Functional Airspace Blocks (FABs; see later in this chapter) they have to conclude an arrangement on

---

10   Point 6(1) of the Explanatory Memorandum of the framework regulation – COM(2001) 123 final/2 of 30 November 2001, and Art. 1(4) and (5) of the framework regulation.

11   www.gibnet.com/texts/trip_1.htm.

12   The proposals were made in two stages during 2001 and were subsequently revised on linguistic grounds.

13   Regulations are directly applicable and require compliance by all parties affected (public as well as private). Directives, on the other hand, would be binding only on the states, not on private parties, and require that the state adopt internal legislation to give effect to the obligations set out in the Directive and to ensure compliance by private parties.

14   As the European Aviation Safety Agency develops, consideration will be given to entrusting that organisation also with tasks in the field of air traffic management.

15   Framework regulation, Art. 4.

the supervision for the service provider managing that block.[16] More generally, there is no need for member states to set up individual authorities for each state but they are at liberty to set up cooperative structures for that purpose, which may help states to develop effective bodies in the face of scarce local resources.

The legislation gives the Commission extensive powers to elaborate detailed rules to give effect to the principles set out in the legislation. These 'implementing powers' are exercised in accordance with the 'comitology' principles[17] through constant interaction with member states, by means of the Single Sky Committee. It should be noted that the Committee itself does not take decisions but issues an opinion on a Commission draft. The Commission cannot proceed to adopt the measure unless it has been supported by a qualified majority of member states.

An unusual feature of the Single Sky Committee is the presence of two representatives (rather than a single representative) for each member state[18] – this allows states, who so wish, to delegate a member of both their civil and military ATM communities, thereby making it possible for the military to be closely involved in the elaboration of Community legislation in this area. In addition, the association of civil and military stakeholders will help member states to organise structures for coordination at the national level, and, if necessary, arbitrate between these two communities, that too often remain separate.

The Rules of Procedure of the Committee allow for the participation of EUROCONTROL because of the specific expertise and pan-European scope of that organisation, and of third countries that have agreements with the EU associating them with its aviation policy (at the time of writing: Norway, Liechtenstein, Iceland and Switzerland; in 2006, the EU concluded an agreement to create a European Common Aviation Area[19] and to extend this approach to eight South-East European partners.[20] Further countries with which aviation relations are being discussed include Ukraine and Morocco). Romania and Bulgaria joined the EU on 1 January, 2007, and there is a longer term perspective of membership for Turkey. As a result of these aviation agreements, it can be expected that the Single Sky legislation will eventually apply to most of geographical Europe. The current agreements cover all of Western Europe, and further agreements may be expected with Eastern European and North African neighbouring states.[21]

---

16   Service provision regulation, Art. 2(3).

17   Council Decision No 1999/468 of 28 June 1999 laying down the procedures for the exercise of implementing powers conferred on the Commission, OJ 1999, L 184.

18   Framework regulation, Art. 5.

19   The European Common Aviation Agreement was signed by the EU and by the partner countries in June 2006 and is in the process of ratification.

20   Albania, Bosnia and Herzegovina, Bulgaria, Croatia, the Former Yugoslav Republic of Macedonia, Romania, Serbia and Montenegro, and the United Nations Mission in Kosovo.

21   It should be noted at this point that the geographical scope of the Single Sky legislation as regards EU member states is not explicitly defined. The guiding principle is that it applies not only to the territory *stricto sensu* of member states, but also to other areas for which they have the responsibility (essentially high seas areas entrusted to states by ICAO).

Building on the positive experience of the industry and social group in 2000, the legislation sets up an Industry Consultation Body, composed of high-level representatives of ANSPs, airspace users, airports, manufacturers, and professional staff representative bodies, that will advise the Commission on technical aspects of the implementation of the Single Sky.[22] The Commission expects this body to provide the framework for the development of a consensus among the various stakeholders on the technological choices that need to be made in this field and on the timetable to be followed for the introduction of new equipment and procedures. The Industry Consultation Body for example concentrated on a work programme in the field of interoperability; it also issued opinions on important regulatory initiatives such as the SESAR implementation (see later) and air navigation charges.

The maintenance of monopoly arrangements for air traffic services implies that one cannot rely on market mechanisms to ensure dynamic service provision, customer responsiveness and emphasis on cost containment. Therefore, the legislation introduces systematic performance review as an instrument to identify best practices and to organise their dissemination – Framework regulation, Art. 11: this provision refers to the contribution to be made by EUROCONTROL's Performance Review Commission which, over the years, has published a series of very valuable annual and topical reports.

7.3   EUROCONTROL in the Single European Sky

Whilst initially, there may have been some concern about the possibility to organise the coexistence of the EU and EUROCONTROL, the Single Sky package has found a balanced way to develop synergies between the two partners and to concentrate on what each of these organisations does best. The institutional set-up of the Single European Sky acknowledges the contribution which EUROCONTROL can make and sets out a number of mechanisms for that purpose. As mentioned, the European Community has joined EUROCONTROL as a full member – this ratification process is still under way and will make it possible for the Commission to reflect the common objectives of EU states within EUROCONTROL and to make sure that the political EU agenda is supported by, and consistent with, work on technical and operational issues. The signing of the Memorandum of Cooperation (December, 2003) between the European Commission and EUROCONTROL, provided a platform for the contribution by EUROCONTROL to the Single European Sky initiative. This Memorandum leads to the definition of a joint work programme that organises technical input into future EU legislation. Most of the implementing rules which the Commission will have to adopt, will be based on technical input to be provided by EUROCONTROL under a system of 'mandates' that define the objectives and timetable of these rules.[23] When EUROCONTROL accepts

---

22   Framework regulation, Art. 6 and 10.
23   Framework regulation, Art. 8.

these mandates, it will undertake to prepare a proposal, conduct an extensive and transparent consultation process, and deliver the results to the Commission for adoption after discussion in the Single Sky Committee. This approach opens the way for the convergence of the political agenda with technical work, in a structured manner.

EUROCONTROL's broad membership, covering essentially all of geographic Europe, makes it possible to extend the Single Sky beyond EU member states and countries that are linked with the EU through aviation agreements, so as to reach countries in the European periphery such as Moldova and Armenia. EUROCONTROL's work under the mandates will find its way into the ATM environment in those countries and, even where it does not have a solid regulatory basis, contribute to the creation of a seamless pan-European airspace.

7.3.1   The Single European Sky Military Dimension

One of the most challenging aspects of the Single Sky is the involvement of the military in this initiative. Representatives of air forces and military ATM have worked closely with the Commission to develop the Single European Sky initiative over the past years, and demonstrated a genuine interest and willingness on their part to play a role in this process. However, it has not been easy to move from policy discussions to regulatory measures. The Single Sky remains rooted in the traditional 'first pillar' of the EU, the European Community, which focuses on economic issues, including transport. The European Community has traditionally had difficulties accommodating military subjects within its activities, even though there is no blanket exemption for these matters (Koutrakos, 2000). While the EU's more recent 'second pillar', the Common Foreign and Security Policy, provides some opportunities to develop action on military subjects, it has been difficult to do so in the field of air traffic management, for which there was no consensus (within this second pillar) to consider it as a priority area for cooperation. The pragmatic approach which followed, was to elaborate the civil aspects of the Single Sky and to concentrate on the interfaces with the military, rather than to put in place a military approach. Thus the problem was to define the scope of the legislation as regards the military, and to develop the right structures for channelling input by the military into the Community process.

First of all, the Single Sky legislation makes it clear that it does not apply to military activities *per se*, defined as 'military operations and training'.[24] In the same sense, the objective of the initiative is defined as enhancing safety and efficiency for GAT,[25] thereby excluding OAT, i.e. flights operating in accordance with military air traffic service procedures, as opposed to movements of civil and state aircraft that are carried out in conformity with ICAO procedures. As a 'belt and braces' measure the legislation contains a safeguard provision in order to protect essential security and defence policy interests.[26] As a result, the military

---

24   Framework regulation, Art. 1(2).
25   Framework regulation, Art. 1(1).
26   Framework regulation, Art. 13.

are not constrained in their ability to conduct operations of a military nature. However, that does not imply that military flights and aircraft are outside the scope of the EC legislation; unless they benefit from an explicit exemption, aircraft equipage and flights transiting civil airspace will have to comply with applicable requirements. In addition, as mentioned, some military organisations provide services to civilian users and these will have to operate in accordance with rules on service provision.[27]

Therefore it is desirable that the military be full partners in the development of any future EU rules that may affect them as airspace users or as providers of ATM services. Any legislation will need to reflect the requirements and constraints of the military in the same way as those of civilian use. This is reflected in their participation in the Single Sky Committee referred to above and, more generally, in their involvement in the drafting of implementing rules by the Commission or by EUROCONTROL on a mandate from the Commission. It is understood, however, that it may be necessary to make specific derogations for the military if the application of the normal rules stands in the way of the proper conduct of defence and training missions but, hopefully, the early involvement of the military in the drafting of these rules will obviate the need to rely on this type of safeguard.

Secondly, member states have agreed to include the subject of the Flexible Use of Airspace (FUA) concept in the Single Sky package. This creates possibilities for the development of clear and enforceable rules on the sharing of airspace between civil and military use and on the interfaces to be developed for that purpose, and for the reinforcement of the harmonised application of this important concept. While it is, of course, problematic to do so without trespassing at all on military operations and training, both civil and military air traffic management communities are committed to pursuing that approach. A regulation on FUA was one of the first implementing rules to be adopted by the Commission following the entry into force of the Single Sky legislation.[28] It establishes rules and procedures between civil and military authorities responsible for ATM. This will increase safety and the efficiency of aircraft operations by ensuring the best use of the available airspace. In this way, the airspace can be managed as a continuum in which the requirements of all users – civil and military – can be accommodated.

In addition, the statement made by the member states on military issues related to the Single European Sky[29] reflects their commitment to work together in areas not covered by the Single Sky legislation, to support work in this area. As a result, we can make progress outside the formal Community context, e.g. towards the development of harmonised rules on OAT, joint exercises and, possibly, joint training areas. This work will no doubt create a common interest that may eventually lead to further-reaching measures, all the more since the draft

---

27   Service provision regulation, Art. 7(5).

28   Commission regulation (EC) No 2150/2005 of 23 December 2005 laying down common rules for the flexible use of airspace, OJ 2002, L 342/20.

29   See footnote 1.

treaty establishing a Constitution for Europe integrates the three EU 'pillars' and will thereby facilitate the accommodation of military subjects within traditional Community policies.

7.3.2   Defragmenting the Airspace and Consolidating Air Navigation Services

The high-level group report had identified fragmentation as one of the main causes for the disappointing performance of ATM in Europe. As a result, the Single Sky legislation sets out a number of mechanisms that are intended to facilitate the restructuring of airspace and of ANSPs, with a view to creating a more integrated, seamless environment.

Without introducing an actual market environment, the service provision regulation makes it possible for ANSPs to operate outside their home state. The regulation introduces a certification mechanism, to be administered by national supervisory authorities, that ensures compliance with a set of common requirements relating to technical competence, financial strength and management of ANSPs.[30] Consequently, the certification mechanism creates a harmonised baseline within the EU that elaborates on the ICAO principles and provides assurances that service providers continue to meet public interest requirements even when the trend towards corporatisation and privatisation continues.

In addition, ANSPs that benefit from a certificate are entitled to offer their services to other ANSPs, airspace users and airports within the EU.[31] This does not mean that ANSPs will automatically be in a position to provide these services outside their home state, but it does create an option for users of air navigation services to procure these services from a supplier elsewhere in the EU, rather than to procure them from a domestic supplier or to supply them by their own means.[32]

Details of the certification process for different categories of ANSPs were laid down in an implementing rule.[33] This 'common requirements' regulation establishes a set of rules against which ANSPs in Europe will be certified. These requirements include technical and operational competence, financial strength, organisational structure, human resources, liability and insurance cover, reporting systems, and processes for safety and quality management. In the area of safety, the regulation incorporates the relevant EUROCONTROL Safety Regulatory Requirements. Certificates will be valid throughout Europe due to the principle of mutual recognition and will thus contribute to a greater cooperation between service providers and to more cross-border services.

----

30   Service provision regulation, Art. 6.

31   Service provision regulation, Art. 7(6) and (8).

32   It should be kept in mind that for the core air traffic services, member states are entitled to maintain monopoly arrangements. However, they are at liberty to designate a provider from another member state rather than continuing the historic reliance on the domestic provider.

33   Commission Regulation (EC) No 2096/2005 of 20 December 2005 laying down common requirements for the provision of air navigation services, OJ 2005, L 335/13.

As a result, the certification mechanism creates considerable opportunities for the provision of cross-border air navigation services and rationalisation of the industry. In order to increase the momentum towards this development, the airspace regulation introduces the concept of Functional Airspace Blocks (FABs).

This notion refers to the fact that ANSPs should be given responsibility for areas that reflect operational requirements to enable optimum use of airspace, taking into account traffic flows rather than reflecting national boundaries. Member states are now obliged to reconfigure their upper airspaces into these FABs. This obligation currently only relates to airspace above 28 500 feet, but lower airspace may, of course, also be included, which may often be necessary in view of the close link between upper and lower airspace (see Chapter 1). This duty will entail the critical review of existing arrangements for airspace organisation and their gradual adjustment to more functional areas, moving away from the current focus on national territories and leading towards the creation of cross-border blocks along the lines of the current Maastricht Upper Area Control Centre (Maastricht UAC).[34] When a cross-border FAB is created, participating member states have to agree on the designation of one, or several, air traffic service providers for the area.[35]

The combination of the certification mechanism and of the establishment of FABs opens the way towards the consolidation of service provision, whereby larger cross-border areas can be entrusted to one, or a combination of, service providers to be managed in an integrated manner. This process also creates opportunities for the rationalisation of current operational arrangements, in particular through the combination of small air traffic control centres into optimised units. Some of these changes will require personnel moves, and with a view, *inter alia*, to organising the mobility of air traffic controllers, the legislator agreed to the creation of a Community air traffic controller licence, in a directive[36] which harmonises the conditions for access to the profession, thereby increasing safety standards. The mutual recognition provisions facilitate mobility of controllers in the context of cross-border air traffic control centres. The social consequences of this are discussed in Chapter 8.

The shift towards FABs was one of the more critical elements of the political discussions. Some member states viewed the establishment of cross-border control zones as impinging on their sovereignty. The Commission took the view that member states retain responsibility for their airspaces under the Chicago Convention, but that nothing prevents them from exercising this responsibility collectively under the EU umbrella and from agreeing uniform rules. European Parliament was also keen to define mechanisms to ensure rapid progress towards

---

34   With a view to stimulating this review process, the Commission has mandated EUROCONTROL to carry out a number of activities in order to identify practical issues arising from this complex exercise and to develop common solutions.

35   Service provision regulation, Art. 8(4).

36   Directive 2006/23/EC of the European Parliament and of the Council of 5 April 2006 on a Community Air Traffic Controller Licence, OJ 2006, L 114/22.

optimised cross-border airspace arrangements. Eventually, a compromise was reached around the confirmation of member states' sovereignty over their airspace[37] and the establishment of a so-called 'bottom-up' process that conditions the creation of FABs on the consent of all member states concerned rather than an EU or EUROCONTROL organised 'top-down' approach. On the insistence of Parliament, fearing that this approach would consolidate the *status quo*, however, the Commission undertook to evaluate the results of the bottom-up process by the end of 2008 and to make proposals for amendment if necessary.[38]

This perspective has encouraged member states and service providers throughout the EU to explore possibilities for setting up FABs. The Commission evaluated the progress made in early 2007 and found that member states and ANSPs should step up their efforts to reduce fragmentation, to increase service levels and to achieve cost reductions;[39] it identified priorities for further action, with a view to making substantial progress towards the creation of these FABs by the end of 2008, when it would evaluate the situation again.

Airspace organisation under the Single Sky legislation entails a number of different initiatives with a view to promoting the integration of member states' airspaces into a seamless structure. The airspace regulation envisages that a European Upper Information Region be created (see also Chapter 1), for which aeronautical information shall be available in a central location. Work is underway towards a formal discussion with ICAO on the creation of such a region and its inclusion in the global aviation framework. Airspace classification also needs to be harmonised around a limited number of categories with corresponding access opportunities for aircraft operators, so that airspace and route planning can become simpler and more straightforward (see also Chapter 2). A Commission Regulation introduces a common classification for all airspace above 19 500 feet and clear rules for access to this airspace. This creates a transparent framework for flights operating over European borders and facilitates access for VFR flights (see Chapter 1). The rule will also make the airspace system more understandable for pilots who may not be familiar with local conditions.[40] Finally, the airspace regulation also contemplates the adoption of implementing rules on ATM and the development of common principles for route and sector design. The entry into force of these airspace measures will facilitate the management of air traffic across national borders and the setting up of FABs.

---

37   Framework regulation, Art. 1(2).

38   Commission statement on the process for the establishment of functional airspace blocks, see footnote 1.

39   Communication from the Commission to the Council and the European Parliament: Building the Single European Sky through functional airspace blocks – a mid-term status report, COM(2007) 101 final.

40   Commission Regulation (EC) No 730/2006 of 11 May 2006, OJ 2006, L 128/3.

7.3.3   Air Navigation Charges

Perhaps one area in which there is need for further development of the industry relates to the rules on air navigation charges. The current system, based on cost recovery, is a sensible approach in order to avoid any significant discontinuity that might jeopardise the financial basis for air navigation services. In the long term, however, this mechanism is increasingly challenged because, in a monopoly context, it does not in itself create any incentives for cost reduction and because it shifts the burden of traffic downturns entirely to airspace users (in times of crisis, airlines end up paying relatively more for the services they receive when they can least afford it). Earlier attempts to reform this approach have met with limited success. The service provision regulation envisages Community provisions on air navigation charges that to a large extent reflect existing principles and methods – even though there are a number of innovations such as the systematic inclusion of terminal charges, increased transparency and room for incentives for ANSPs as well as for airspace users. An implementing rule was adopted at the end of 2006 that sets out the details of such a charging system, largely mirroring current practice within EUROCONTROL.[41] The existence of Community legislation in this area provides a channel for further discussions on this subject which, over time, could lead to improvements of the funding of the industry.

7.4   SESAR – Modernisation of ATM Systems in the Single European Sky

A more seamless environment also requires additional emphasis on the standardisation of equipment and procedures, which is achieved by the interoperability regulation.[42] The legislation is based on a set of essential requirements relating in particular to interoperability, safety and performance levels, which may be refined by means of binding interoperability implementing rules to be adopted by the Commission for critical components, or for cases where the interfaces have to be defined with great precision. For other items, the regulation follows the 'new approach' – relying on voluntary standardisation initiatives, by means of Community specifications elaborated either by European standardisation bodies[43] or by EUROCONTROL on matters of operational coordination between ANSPs. This dual approach – combining binding and voluntary measures – provides a flexible framework under which EU standardisation ensures interoperability of systems and European industry benefits from a platform on which to develop globally competitive equipment.

---

41   Commission Regulation (EC) No 1794/2006 of 6 December 2006 laying down a common charging scheme for air navigation services, OJ 2006, L 341, 3.

42   See footnote 1.

43   Such as: CEN (European Committee for Standardisation); CENELEC (European Committee for Electrotechnical Standardisation); and, ETSI (European Telecommunications Standards Institute) – in cooperation with EUROCAE (European Organisation for Civil Aviation Equipment).

Interoperability implementing rules on coordination and transfer, and the Initial Flight Plan, were adopted by the Commission in 2006,[44] to be followed by a rule on the Flight Message Transfer Protocol. EUROCONTROL is preparing draft rules on Air-Ground Voice Channel Spacing, datalink and Aeronautical Data Integrity, and starting work on Surveillance Requirements and on Mode S Interrogator Code Allocation. After consultation of the Industry Consultation Body, the Commission has identified priorities for specific mandates to CEN/CENELEC/ETSI (for the development of European standards in cooperation with EUROCAE) and to EUROCONTROL.[45]

The mechanism of implementing rules also may be used to facilitate the coordinated introduction of new technologies. The Commission may elect to adopt requirements obliging manufacturers of equipment, airspace users, airports and ANSPs, to comply with certain indications as to which type of equipment and procedures they should use, as well as to lay down a timeframe for compliance. On this basis, the Commission may prescribe the use of specific technologies and mandate equipage of aircraft and air traffic control centres throughout the Community. This power is particularly significant in the light of the traditional difficulty of agreeing on technological choices, which has considerably delayed the introduction of new equipment and stood in the way of a uniform technical environment (see, for example, the discussion of P-RNAV in Chapter 1).

Obviously, the Commission is not in a comfortable position to make these choices, even where it is supported by EUROCONTROL under the mandate system. Therefore, it intends to rely on the consensus which should emerge from discussions within the Industry Consultation Body. Manufacturers have understood the opportunities resulting from this approach and have prepared an initiative to organise the development and introduction of new technology under a major project: SESAR (formerly known as SESAME). This will develop the next generation ATM systems, with a 2020 horizon. The SESAR initiative combines technological, economic and regulatory aspects, using the Single Sky legislation, so as to synchronise the implementation of new equipment, from a geographical standpoint in all European Union member states, as well as from an operational standpoint by ensuring that aircraft equipage is consistent with ground technological evolutions.

The first phase of SESAR, called the 'definition phase' was launched in early 2006 and put under EUROCONTROL's overall responsibility, but heavily relying on industry input organised by means of a comprehensive consortium. This €60m project is co-funded by the European Commission (under the Trans

---

44   EC Regulations of 6 July 2006 laying down the requirements for automatic systems for the exchange of flight data for the purpose of notification, coordination and transfer of flights between air traffic control units, and of 4 July 2006 laying down the requirements on procedures for flight plans in the pre-flight phase for the Single European Sky.

45   Request to EUROCONTROL for the development of specifications concerning IFPS (Integrated Initial Flight Plan Processing System) users manual, ADEXP (Air Traffic Services Data Exchange Presentation), OLDI (On-Line Data interchange) and AMHS (Air Traffic Services Message Handling System).

European Networks budget) and by EUROCONTROL. The definition phase will end in 2008 and will be undertaken by a consortium uniting forces from the whole aviation community. It will deliver an ATM Master Plan, which will define a common goal and vision for the development of the European air traffic control infrastructure together with a precise timetable for its implementation.

In early 2007, the Commission reported on the state of progress of the definition phase and on the preparation of the further work under this project.[46] The second phase of SESAR will be a development and deployment phase, based upon the results of the definition phase, and will organise the next generation of air traffic control systems and synchronise their deployment and implementation. The development phase (2008-2013) will produce the required new generation of technological systems and components as defined in the definition phase. It will be managed under a new governance scheme grouping different stakeholders and institutions within a joint undertaking.[47] The budget for this phase is estimated at €2.3–2.7 billion from the Community, EUROCONTROL and industry. The following deployment phase, through 2020, will be carried out under the responsibility of the industry, without further public finding.

The Commission has put considerable weight on an approach that not only gives responsibility to industry for its own future, but is also open to participants from outside Europe. European airlines operate on a worldwide basis and need technological solutions that are not specific to any region. The ATM Master Plan should be consistent with global planning by ICAO but could also provide a point of reference for other regions in the world that need to modernise their ATM infrastructure, thereby also providing commercial opportunities to European manufacturers. Other technical considerations regarding the future European ATM system, for example the concept of 'free flight' and the delegation of separation, were discussed in Chapter 3. The potential for inclusion of environmental and socio-economic impact assessment in the ATM Master Plan, will be discussed in Chapter 8.

7.5   Conclusions

The Single Sky process is still relatively young, and much work still needs to be done. From the Commission's perspective, the main priority is, first of all, the completion of the regulatory framework. Member states and service providers are expected to make substantial progress towards the creation of FABs – the Commission expects to see concrete realisations by the time its assessment of the bottom-up process is due, at the end of 2008. All stakeholders should focus

---

46   Communication from the Commission: State of progress with the project to implement the new generation European air traffic management system (SESAR), COM(2007) 103 final.

47   Council Regulation (EC) No 219/2007 of 27 February 2007on the establishment of a Joint Undertaking to develop the new generation European air traffic management system (SESAR), OJ 2007, L 64, 1.

on the collaborative endeavour to define and implement next-generation ATM systems through the SESAR initiative.

The Commission will submit a first report on the Single Sky implementation in autumn 2007. With a view to assessing the overall situation of ATM in Europe and the contribution that the EU legislation has made, the Commission requested EUROCONTROL's Performance Review Commission to carry out an evaluation (EUROCONTROL, 2006d). The Performance Review Commission found positive effects of cooperation between ANSPs, and between states, and welcomed the clearer separation between ANSPs and regulators. On the other hand, it wished for a clearer framework for performance improvements and formulated suggestions to improve the regulatory capacity of NSAs, to streamline the process of creating FABs and stimulating cross-border service provision, to intensify civil-military cooperation, and to ensure that new legislation was effective, yet flexible.

Adjustments to the regulatory framework will no doubt be made in proper time. A number of items are already identified in the legislation to be revisited in a few years' time: the application of the airspace regulation to lower airspace is a priority, as there is probably even greater need for coordination of airspace use for lower than for upper airspace. The provisions on Flexible Use of Airspace and the member states' statement on military issues open the way for closer involvement of the military in the Single European Sky initiative, and as cooperation in this field intensifies, there may be scope for additional actions in this area.

On the institutional level, perhaps the greatest challenge will be for member states to set up effective national supervisory authorities. For many of them it will require considerable effort to build an effective structure at arms length from the service provider. With the development of cross-border services it may perhaps be preferable to pursue the development of supervisory authorities on a regional level, either to reflect the structure of FABs, or in connection with the eventual broadening of the powers of the European Aviation Safety Authority to the field of air traffic services.[48]

No doubt the development of the EU's regulatory powers and the progress of ANSPs towards stronger cross-border organisations will entail some consequences for EUROCONTROL. In particular, the prospect of an involvement of the European Aviation Safety Agency in this field may lead to a reflection on the role of EUROCONTROL as an intergovernmental organisation.

Finally, air traffic management, as other aviation activities, is quintessentially global. The EU will need to embed its action in this worldwide framework by active participation in the ICAO process, and to coordinate it with the main partner countries outside the EU.

Thanks in particular to careful preparation, broad interest from stakeholders, high-level political support and transparent working methods, the Single European Sky initiative has made rapid progress. In the end, however, the initiative will be judged on its success in improving the performance of air traffic management for the benefit of airspace users and the travelling public.

---

48   Regulation No 1592/2002 of the European Parliament and of the Council of 15 July 2002; OJ 2002, L 240, recital 23

Chapter 8

ATM and Society – Demands and Expectations

Nadine Pilon
EUROCONTROL Experimental Centre

8.1 Growth and Challenges for Air Transport

Change in air traffic management (ATM) can be addressed from many different perspectives. In this chapter, we have chosen to focus on society's demands regarding the air transport sector and how this affects air traffic management.

The world of air transport is changing, not only though the evolution of aeronautical technologies and economics, but also through the interaction of society with ATM – which the industry must also take into account in terms of future strategies.

The first part of this chapter will address society needs, capabilities and expectations, and practices likely to affect air transport's evolution. The second section will focus on societal factors of change in air transport, and the third section will show how the ATM world is responding to these challenges. Finally, we will discuss what else is needed, in particular in what can be called the 'relationship' between air traffic management and society, in order for the industry to keep pace with society's expectations.

The *raison d'être* of the air transport business is society's need for mobility, both for passengers and for freight. Air transport provides mobility to citizens and goods with a continuing growth in demand, which is an indicator of the success of the sector in general in the eyes of society – even if the benefits (and difficulties) are not equally distributed.

As we explored in Chapter 1, commercial air transport is a relatively young industry, experiencing continuous growth rates since its beginning in the middle of the twentieth century. Despite worldwide shocks such as oil crises, Gulf wars and terrorist attacks, growth ultimately returns to around 3–5 per cent per annum, and current forecasts indicate that this trend will continue over the next 20 years, in most scenarios for the future (EUROCONTROL, 2004a; see also Chapter 6). This growth is the result of the desire of an increasing population to travel long distances quickly, and of economic growth. International in essence, it fosters globalisation of business and economies. It increases the mobility of citizens and accessibility to otherwise relatively inaccessible regions. It brings wealth, growth and business opportunities to the economy (ACARE, 2003; ACI, 2004b; Cooper, 2005; ATAG, 2005).

In this section, we will briefly address some of the challenges brought about by the growth of traffic on the future evolution of air transport, factors such as: the negative impacts of air transport and the resistance they trigger in society, the emergence of air transport capacity issues and the sector's dependency on reasonably-priced fuel – the rise in oil prices (and other charges) that may affect ticket prices, and influence the economics of air transport.

8.1.1   The Capacity Challenge

The growth in demand for air transport has generated new challenges for capacity and safety, many of which we have explored in Chapter 3. In response, manufacturers develop new types of aircraft, airlines open new routes and adapt their fleet. On the infrastructure side (ATM and airports), it also leads to the need for further investments in airport extension and ATM modernisation.

In response, the ATM world develops collective strategies to increase airspace capacity while maintaining safety. A significant example is the introduction of Reduced Vertical Separation Minima (RVSM) coordinated through EUROCONTROL: a Europe-wide operational measure to increase airspace capacity by safely reducing the separation between airborne aircraft, implemented on 24 January 2002 (overnight) in all EUROCONTROL member states.

Another significant initiative is the Single European Sky (SES) legislation designed by the European Commission to reduce airspace fragmentation in Europe and adopted by the European Parliament in April 2004. This initiative carries the potential to deeply restructure the provision of air navigation in Europe. Examples of areas of change brought by the Single European Sky legislation include the reorganisation of air navigation service provision into Functional Airspace Blocks, independent of national boundaries, and the creation of a Community air traffic controller licence in place of the national licences.

As a complement to the Single European Sky initiative, the SESAR project (SESAR Consortium, 2006a; see also Chapter 7), a major pan-European industry-led collective programme for the modernisation of the air traffic management infrastructure, supports the SES, in particular with its technical objectives of systems interoperability and capacity enhancement. It aims to define and implement a new air traffic management concept of operations able to overcome current capacity issues: deployment is planned from 2013, in addition to fostering research for the period after that.

Through these initiatives (which we discussed in more detail in Chapter 7), it is evident that European ATM is undergoing a process of change in several of its fundamental aspects: operational, technological and institutional.

8.1.2   The Energy Challenge

The air transport sector is highly dependent on the availability of oil at a reasonable price. The end of cheap oil is frequently being predicted. Whilst it is quite difficult to predict when oil supplies will no longer be available for aircraft, the Hubbert Peak Theory (Hubbert, 1956) indicates that the price of oil will inevitably rise when demand overtakes the production capacity (the rise in oil prices observed in 2005–2006 might be the beginning of this trend, despite subsequent falls in late 2006). Whenever such rises actually occur, airlines will be the first to be hit, and they will have to adapt to the end of cheap oil.

At the same time, consumers' habits and priorities might be affected in all aspects of their lives by more expensive oil, and the demand for air transport may change. Similarly, other factors, independent of air transport, may affect both leisure and business demand – wars and conflicts, terrorism, and pandemics (e.g. avian influenza). All air transport stakeholders are likely to be affected directly, or indirectly, by such trends.

The issue of oil consumption in air transport is intimately connected with the issue of gaseous emissions, which is also progressively having an economic impact on the air transport industry *via*, in particular, the introduction of the Emissions Trading Scheme. The industry is already involved in efforts to reduce both fuel burn and gaseous emissions, and, as analysed in (Metrot, 2006), will increasingly need coordinated actions to improve its energy efficiency, in a perspective of overall sustainability.

8.1.3   The Sustainability Challenge

Whilst bringing considerable benefits to the economy and society, air transport also brings negative impacts to society and the environment. Therefore, whilst this growth is beneficial to business and industry, at the same time it raises increasing concerns, particularly in some parts of Europe, in addition to raising the issue of energy availability – in the medium- to long-term. The (impossible?) challenge for air transport sustainability is to strike a balance between these social, economic and environmental impacts.

Several definitions of sustainability exist. The purest definition (Bruntland, 1987) includes the requirement that the activities of the present generation should not inhibit the capacity of future generations to perform similar activities.[1] Applied to air transport, this radical definition is not compatible with the current energy sources used for air transportation nor with the environmental effects of aviation: both on the atmosphere, affecting the Earth's climate (IPCC, 1999), (IPCC, 2001), and also on the local environment around airports, with issues of noise pollution and deterioration of the local air quality (EUROCONTROL Experimental Centre, 2005), as we discussed in Chapter 5. As growth in air traffic increases, so do gaseous emissions and noise nuisance. Fossil fuel is a non-renewable energy source which pollutes both the local air around airports and the higher atmosphere. Opinions differ regarding what should be done, from full confidence in the development of technology that will allow the next generation

---

1   The Bruntland report, 'Our common future', defines sustainability as follows: 'Sustainable development is development that meets the needs of the present without compromising the ability of future generations to meet their own needs'.

of aircraft to use other energy sources, to strong opposition to any future growth of air traffic to ensure that aviation fully contributes to the worldwide efforts agreed in the Kyoto Protocol. The rise in oil prices is further widening the gulf between these two contrasting positions.

The CONSAVE project (CONSAVE, 2005) was designed to integrate environmental impacts into future scenarios. In its most stringent scenario, it envisages a possible stagnation of air traffic growth due to strong environmental constraints. Our aim here is not to discuss whether the growth in demand is going to continue in the future, or whether or not it should be regulated, but rather to see how this growth is changing the industry, in particular through shifts in society's perception.

Achieving air transport sustainability involves striking a balance between the social, economic and environmental effects of air transport on society. As explored in Chapter 5, for a global business like aviation, this means responding to local and global effects both in the short- and long-term. The term 'glocal' (Swyngedouw, 2000) encapsulates the challenge faced by air transport (as other global industrial sectors) to build an acceptable compromise with their local neighbouring communities whilst not losing sight of the global business requirements.

Although in countries such as the UK and Sweden, aviation receives high media attention, in most countries in Europe, air transport is not as exposed to public scrutiny as are other industries, such as the food, agriculture and nuclear industries. Although increasingly 'tainted', especially in terms of the emergence low-cost carriers, climate change coverage, and lengthy security procedures, air transport still has a positive image of progress, luxury and power, which does much to protect the industry against the growing opposition of certain populations which bear the brunt of the associated nuisances and risks. By committing to an ethically responsible way of doing business, a number of industrial sectors have been able to create a new method of governance and to gain a real competitive advantage in a changing world.

8.1.4   *The Intermodality Challenge*

In the context described above, much interest is currently devoted to the integration of air transport with the other modes of transport, either to increase airports' accessibility, or to improve intermodality – for better travel choices for passengers and improved freight distribution. In *Airports as Multimodal Interchange Nodes* (ECMT, 2005), airports are analysed as multimodal hubs in a wider transport network.

At the political level, the Lisbon Strategy for Europe put mobility at the heart of European competitiveness for the EU (Kok, 2004). In its 2001 White Paper on European transport policy entitled 'European Transport Policy for 2010: time to decide' (European Commission, 2001a), the European Commission, placing users' needs and competitiveness at the heart of the European transport policy, indicated its willingness to adapt transport to mobility needs and new challenges, and put an emphasis on promoting rail transport – incidentally limiting the promotion of air transport growth – and to foster intermodal transport. In its 2006 mid-term review, the Commission promoted co-modality between all modes of transport to better integrate transport modes for the benefits of European mobility, and emphasized the need to overcome the bottlenecks in infrastructures (European Commission, 2006).

It is with a view to improving access to airports and enhancing intermodal transport for passengers, that such increasing emphasis is being placed on the integration of air transport with other modes of transport. The use of regional airports by new airline operators (in particular low-cost carriers) also increases the need for more connections to other transport modes. In the future, the possible emergence of more 'local' communities with decentralisation of companies and services to regional levels, triggered by a desire (or constraint) to travel less, may also increase the need for integrated and decentralised transport. Enhanced integration of the air transport infrastructure is likely to increase the catchments of airports and to further increase the net demand for air transport in the future, while potentially decreasing the concentration of the various sources of nuisance. (Issues of regionalisation and changing catchments are also explored in Chapter 6.)

Decisions on such infrastructure development projects involve many stakeholders. Local authorities see, in airports and multimodal transport infrastructures, opportunities for economic development in a context of competition between urban areas. Cities also face rules and regulations regarding the sustainability of urban transport, and have a responsibility to their local communities. This increased connectivity between air and other modes of transport may, in the future, drive a level of reconciliation between local short-term social preoccupations and longer term, global sustainability issues.

8.2   Society Factors of Change in Air Transport

Despite the multiple benefits brought by air transport to society in terms of economic growth -wealth, mobility, rapidity, availability and abundance of goods, regional airports stimulating local activity (Bråthen et al., 2006), particularly in less mature European economies such as Romania and Bulgaria – citizens are increasingly reluctant to accept the risks that goes with it. In order to have continued growth to meet demand, together with many other challenges in technology, infrastructure, operations, regulations, financial and economic issues, air transport stakeholders already have to also address society's concerns: safety risks for passengers and for third parties; health, noise and pollution, congestion, economic depreciation (which may also occur in the area around an airport), and security – which, with more time spent at airports, could promote modal shift to rail. Added to these factors, we have human rights issues and the global risks of climate change.

The resistance triggered by the negative impacts of air transport among the population, especially among those who live in the vicinity of airports and under flight paths (even at more than 50 km from airports as was the case with the reorganisation of the arrival and departure routes over Paris) are factors likely to affect the sustainability of air transport in the future. As Faburel explains (Faburel, 2005), objectives related to airports and air mobility needs are being increasingly debated in society.

The air transport industry, like other mobility sectors, faces the challenge of building a constructive dialogue with stakeholders, whose interests are not always compatible with the concept of sustainable development. The air transport industry must recognise this growing need and understand the direction in which further work is required to consolidate the level of social acceptability needed for (sustainable) development. Sometimes, this dialogue is focused on the local region and urban areas close to airports; however, when national economies are in question, air transport growth may become (increasingly) a subject for political debate, such as in the UK or in Sweden, as analysed in a study submitted to the 2007 European Transport Conference (Mahaud and Courty, 2007).

When trying to analyse the underlying factors in European society which explain these evolutions, a few trends seem to appear as factors of change in air transport and air traffic management:

- society sensitivity to risks;
- change in political decision-making processes;
- citizens' participation around airports.

These society factors need to be understood and integrated into the strategies of the air transport industry, including ATM, when aiming at contributing to a greater sustainability of air transport.

8.2.1   Societal Sensitivity to Risks

Aversion to risk seems to drive many societal developments: in politics, regulation, and jurisprudence. The risk of terrorism is reinforcing this aversion. The philosopher Paul Virilio speaks of the 'cold panic' that would follow the 'cold war' after the 11 September 2001 attacks. In the 'society of risk', as referred to by Beck (Beck, 1992), analysing how the emergence of industrial risk has affected modern societies, citizens increasingly sensitive to risks generated by industry would pay increased attention to decisions affecting industry developments, regulation and impacts. The growth of air traffic increases the perception of risk in the public:

- risk of flying: current forecasts announce the doubling of traffic globally in Europe by 2025 (EUROCONTROL, 2006e). This predicted growth of flights may produce more incidents in absolute numbers, and increases the risk of collision. In order to maintain its adequate level of safety, the air transport industry, and, in particular, ATM, implements continuous reinforcements of safety assurances, but this has little repercussion on the *perception* of risk;
- risk of being attacked: airplane hijackings have always triggered a lot of public attention. But the 11 September 2001 attacks have created in the public imagination the possibility that aviation can potentially have a mostly negative

image. In addition, the burden of long security checks at airports increases the perception of security risk associated with air transport;

- risk of living around airports: different kinds of consequences can be associated with living in the vicinity of airports, some positive (economic, accessibility, tourism) and others more negative, such as those shown in Table 8.1.

Table 8.1   **Possible negative consequences of living near an airport**

Consequence	Description
Effects of air traffic on surrounding populations	Aircraft noise and local air quality pollution, due to vehicles and power supplies, which may affect health, quality of life and socio-economic characteristics of local communities, as has been analysed in several existing publications (Faburel, 2006).
Third party risk	In the case of an accident at an airport, surrounding populations may be exposed to dramatic consequences, such as with the El Al accident in Amsterdam, in 1992, or the Concorde accident in Roissy, in 2000. Such accidents produce important media coverage and a strong impact on public opinion. Even in the absence of accidents, the simple fact of seeing an aircraft getting close to the ground can reinforce the perception of such risk.
Possible pandemics (e.g. avian influenza)	Airports are important entry points inside territories through which the globalisation of exchanges can also promote the spread of viruses.

These risks are inherent with air transport and grow with air traffic volumes, but their perception by the public varies according to different factors, such as economic wealth, lifestyle, and activities in the vicinity of airports. EUROCONTROL-commissioned studies (Bristow et al., 2003; Bristow et al., 2005) have shown, for example, that annoyance around the airport at Bucharest-Otopeni (now Henri Coanda International) was less due to aircraft noise than in Lyon: air traffic growth may be more welcomed in less mature economies, than in maturer economic regions. When air transport and economies develop, the perception of risk seems to progressively become stronger, relative to the benefits to society.

It appears that society could be becoming more sensitive to, and more aware of, air transport safety risks and/or environmental impacts, and possibly becoming less tolerant of operational errors. An illustration is the case of the Linate accident where a runway incursion resulted in a collision, with fatalities, in 2001. Both the judicial investigation and the technical investigation were heavily reported in the press, in particular in the regional newspapers. The trial resulted in eight ATM personnel (from front-line operations to top management) being jailed (and subsequently went to appeal). The way in which this, and a number of similar cases, were reported in the press, and handled in the courts, are analysed in a recent report (Mahaud, 2005) and identifies a lack of support by society for air traffic management. There remains a tendency for these cases to be addressed not only through changes in the system but, additionally, by establishing culpability and punishment, on the basis of personal liability. This poses the question of whether such cases point to a trend where, in modern societies, air transport is considered as a mature industry in which failures of that system are less and less tolerated.

8.2.2   Change in Political Decision-Making Processes

National, political debates on air transport are mainly related to airport creation, expansion or relocation. Therefore, the future development of air transport depends partly on how well airports integrate with their surrounding area. Airports are consequently the focal point of a whole range of conflicting interests and potential disagreements. They are also the first point of contact for the public, as gateways to the air transport system for passengers. They often occupy residential areas and are surrounded by a wide range of economic activities (with investors and workers), as well as local authorities. When airports need to grow to allow the rest of the industry to grow, resistance sometimes successfully halts airport expansion plans.

Of late, only a small number of airport expansion projects have actually been approved in the denser parts of Europe. In the UK for instance, airport expansion projects are being debated at the level of the Parliament through a Governmental White Paper. Even if the public debate on this subject is not as developed in every country in Europe, this does indicate a possible trend, whereby industry plans for growth may encounter increasing resistance from society.

The role the public plays in public decisions is increasing in Europe. This trend is encouraged as part of the new concept of European citizenship, and it stresses the need to provide accurate and digestible information to knowledgeable society players. Therefore, the gap slowly widens with these society players, who are increasingly expecting to be involved in decisions affecting their quality of life, health, safety, and the environment. A good example of the potential blocking power of society players is the case of Brussels airport where DHL, wanting to expand its operations, was ultimately rejected by the local population despite the economic benefits brought by the company. As a result, Brussels DHL moved to Leipzig.

In most places, more local decisions affecting the neighbourhood are discussed, debated and made with greater involvement of citizens. Under EC Directive 2001/42/EC on Strategic Environmental Assessment,[2] socio-economic and environmental impacts have to be assessed and public consultation is required for projects or programmes affecting populations.

---

2   European Commission Directive on the assessment of the effects of certain plans and programmes on the environment.

Quoting Yves Le Bars, former President of ANDRA (the French national agency for dealing with radioactive waste) – analysing public decision-making practices, there are three evolutionary stages in public decision. While some industrial sectors, such as the nuclear industry in some parts of Europe, are already in the third evolutionary stage of the decision-making process, most sectors, including air transport and the ATM world, are either (mostly) in the first or second stage:

- in the first stage, a hangover from World War II when nations needed reconstruction, most decisions were taken by technical specialists: they had the knowledge, expertise and know-how, which constituted their legitimacy;
- in the second stage, the decisions are split from the expertise: decisions are made by decision makers who only consult technical experts to inform their decisions. This is a generalised practice in most industrial sectors;
- in the third stage, civilians are involved in joint decision processes. Such processes require a significant investment in educating the public, and a genuine openness about the outcome. In return, the result can be a satisfying level of acceptance of the decision by the public concerned. Such practices, fostered by the liberalisation of services and the emergence of shareholders'/stakeholders' satisfaction, correspond to a cultural change towards more of an 'inclusion' culture, in line with the concept of corporate social responsibility.

Decisions regarding air transport do not traditionally involve the public, although this is more common in certain countries. However, this situation is progressively evolving under pressure of from local communities around airports (who may protest against both the effects of air transport and their non-involvement in decisions affecting their lives); from European regulations (such as the progressive liberalisation of services); and from the change of ownership of many operators (airports, in particular). These pressures challenge the legitimacy of the traditional decision-makers and introduce stakeholders and/or shareholders as new partners in the decision-making process.

As an example, the evolution of the mobilisation of local communities around the Paris Charles de Gaulle airport has been analysed by Halpern (2006). She shows how the decision-making process for airport expansion projects has progressively involved local communities more, when liberalisation of services in Europe has lowered the legitimacy of public authorities and governments regarding such decisions. As a prime example, in the UK, the question of air transport development is publicly debated in society and the layperson is developing increasingly informed opinions.

8.2.3 Citizens' Participation around Airports

As we have touched upon, public conflicts with the air transport industry are largely concentrated around, and associated with, airports, and this is where compromises are often negotiated. Irrespective of the air transport industry's desire for growth, final decisions about such growth are frequently ultimately made at the airport site itself, where the challenges are focused. Change may be negotiated with, or enforced upon, local members of society, who want to control the local impacts. This is why it is important to understand and recognise both the causes of such conflicts and the mechanisms and instruments which can be used to address them.

The causes of these conflicts are most often related to noise, even if annoyance around airports can also come from non-auditory factors (Stallen and Compagne, 2006). Depending on the culture and socio-economic contexts, different priorities may prevail at different airports. For example, evidence suggests (Cook and Tanner, 2005b) that at Bucharest Henri Coanda, whilst noise and local air quality are taken seriously by the Airport Authority, and monitored by the Local Environment Agency, both were well within the set limits (whereas the management of waste water was a significant, on-going concern), and a complaints culture could not be said to prevail. Compare, in the distinctly mature complaints culture of Britain, noise and congestion are big issues at London Heathrow.

Conflicts between airports and local authorities sometimes involve questions of fairness *between* communities, when distributing both benefits, and nuisances, associated with airports. Local government often plays an important role in searching for solutions, and local politics may defer the national level. Also, depending on ownership, airport owners may be in a position to develop partnerships with the local communities (in many cases, local government has at least a majority shareholding/level of control).

Mechanisms and instruments to address conflicts may be applied at the airport planning phase and, more generally these days, at each airport expansion phase. This includes gathering and sharing information about the impacts of airport(s) (expansion) on the local area, regulating the anticipated effects, and entering into dialogue with affected populations. The 2005 Commission study on land-use planning and management in the EU (INECO Study, 2005) surveys airport planning procedures in the EU; it advocates better integration of airport planning with spatial and land-use plans, 'subject to strategic environmental impact analysis to ensure public information and participation', in compliance with the Environmental Assessment Directive 2001/42/EC.

Sometimes, negotiation or mediation extends to the Terminal Manoeuvring Area (TMA), and Air Navigation Services Providers (ANSPs) are then on the front line. An example was the reorganisation of the arrival and departure routes over Paris, which affected local areas and populations. In such cases, ANSPs can be a key partner in building constructive dialogue by providing useful information to all parties. The question remains – will greater citizen participation help reconcile the global air transport industry's growth aspirations and the population's need to control the expansion of nuisance in the vicinity of airports?

8.3   Society Challenges for ATM

In this section, we will examine the image of air traffic management in the eyes of the general public, and the ATM contribution to two key air transport objectives:

environmental sustainability and safety. Through these examples, we highlight how the air traffic management responses to societal challenges are mainly technical and operational, fitting with the rather technological culture of the profession, but whereby the harmony with society's perception is less clear.

8.3.1 The Image of ATM

The public's image of air traffic management is rather fuzzy. It is certainly generally not known to the public, except on occasions when its actions affect them:

- delays: often the cause of delays experienced by passengers is attributed to air traffic control (ATC) without any more detailed explanation forthcoming. Actual reasons for delays may range from congestion en route or at the destination airport, to weather conditions, delays caused by the airline, or by some other operator. (Delays, and their costs, are discussed in detail in Chapter 4). When expected delays render the departure slot unworkable – for example, due to a night curfew – the flight might even be cancelled. This aspect of flow management (explained in detail in Chapter 2) is rarely explained to the travelling public, and this may create some confusion, and, indeed, a perception among the general public that it is ATC that is blocking the system. In summer 1999, the delays due to the Kosovo Crisis (see also Chapter 7) in former Yugoslavia during the holiday period were so important that they raised media attention, and EUROCONTROL was accused of being responsible for the delays experienced by passengers, with little explanation why normal operations became disrupted;
- strikes: when controllers go on strike, there is, in most cases, a negative impact on air traffic in terms of punctuality, or even cancellation of flights, which is badly perceived by travellers in particular, and the public in general. Disruption caused by a strike can stay in the public mind for some time afterwards, leading to poor impressions of the party involved. Strike action by airline staff has similarly negative consequences, but less often cause network-wide delays in the same way as a controller strike;
- environmental impacts: the negative impacts of air traffic are not associated with ATM in the public's mind, as revealed in a survey of the image of ATM (Kinchin, 2004) and in a recent study of citizens' views about air traffic growth (Cook and Tanner, 2005a). However, as mentioned earlier, modifications of departure and arrival routes are increasingly subject to formal environmental, social and economic impact assessments. (We will return to this subject later, in a discussion on noise reduction). In such instances, much information on the technicalities of ATM is provided to the layman, who may become quite knowledgeable about route design constraints, and subsequently able to question actual operations;
- accidents: a study of the European press' view on air safety (Mahaud, 2004) has shown that since there is little knowledge of air traffic management among the general public (or maybe because of a poor, or inaccurate, public image) ATM can sometimes be negatively portrayed in certain newspapers in connection

with serious safety events. This both reflects, or perhaps forms, public opinion about the profession. The question raised here is how far a better understanding of ATC in society could reduce the risk of a negative public perception inhibiting efforts to ensure a fair assessment of responsibilities.

Through these different facets, the image of ATM looks more negative than positive, and doesn't seem to reflect the real purpose and mission of the air traffic management function. The possible shift of paradigm in public opinion towards less tolerance to errors, already mentioned earlier, is militating in favour of building a stronger image of ATM, based on transparency and professionalism.

8.3.2   Safety, Transparency and 'Just Culture'

Air transport is widely recognised as a safe mode of transport with good safety records – despite some accidents which attract high media attention. ATM, whose contribution is essential to the safety of air traffic, is, on the other hand, not providing at the European level a satisfactory indication of the level of safety it produces, as stated by the Performance Review Commission (EUROCONTROL, 2006a). This apparent lack of transparency in air traffic management safety, although partly due to professional and legal issues, could become a risk for the public image of ATM.

Why is safety reporting necessary? Safety is the main justification, the *raison d'être*, of ATC in the eyes of the public. From the early days of military operations to the development of civil aviation, safety has always been the prime objective of ATC and, subsequently, of ATM. The industry has been successful – with its excellent safety record, the air transport sector, including ATC, has fostered an image of trustworthiness.

However, this success does have a side effect. It is very difficult to measure the safety 'performance' of ATM, since safety is the absence of accidents, and it is hard to measure an absence. Air transport accidents records do not provide enough data for prevention purposes. Furthermore, the air traffic management contribution to air transport accidents is very low (although public perception and media coverage does not always reflect this), rendering the statistical basis even poorer for prevention purposes. Therefore, other sources of information are necessary to continue to improve ATM safety.

That is the reason why reporting on safety occurrences and incidents is required: in some cases, a safety occurrence may indicate that safety has been compromised and therefore lead to improvements. For instance, the EUROCONTROL Safety Regulation Commission (SRC), as part of its set of ESARRs (EUROCONTROL Safety Regulatory Requirements) to be implemented in the national regulatory framework of its member states, has developed ESARR 2 on safety reporting and severity classification (EUROCONTROL, 2000).

Some issues with safety reporting However, this genuine demand for transparency may, in some cases, cause difficulties for air traffic management on account of the safety culture, confidentiality and even legal issues.

Reporting: in organisations such as ATC, team issues play an important role and a strong safety culture is crucial for safety. Efficient safety reporting is based on trust and therefore takes a long time to become fully embedded in the organisation – and can soon be rejected. While in some operational units implementing safety reporting can be seen as a catalyst for reinforcing the safety culture, it can also be detrimental in other instances where it seems to run counter to trust (Fassert, 2005).[3] For instance, if local safety reporting is strong, but informal, care needs to be given to maintain trust during the change towards a more formalised reporting system.

Confidentiality: in many cases, the confidentiality of safety reporting is seen as a pre-requisite if safety reports are to be filed and used for safety improvement within and between organisations. Confidentiality avoids pointing the finger at individuals, units or organisations. Individuals should voluntarily report their own errors and any other dysfunction of the ATM system, and such reports, rendered anonymous, should be made available for safety improvement. When considering the way safety analyses, investigations and improvement measures are carried out, some may see certain confidentiality clauses as being over-restrictive, however. A delicate issue to resolve.

Legal issues: unfortunately, such confidentiality provision, which is good practice for safety management, is, however, at odds with the legal system: in the case of a legal inquiry, any requested data or file will be released to the judicial authority, regardless of any confidentiality agreement. In certain countries in Europe, even when no accident has actually occured, staff may still be prosecuted because of a safety incident.

'Just culture': in response, the ATM community and EUROCONTROL, for instance through the SAFREP task force, are developing and promoting the concept of 'just culture' in air traffic management.

> The SAFREP task force found that punishing air traffic controllers or pilots with fines or license suspension, as well as biased press reports, has led to a reduction in the reporting of incidents and sharing of safety information. It also recognised that the need for a culture that encourages honest reporting is not yet reconciled with the judicial system and legislators. It warns that the situation may get worse if no immediate action is taken.[4]

Addressing the limit between tolerable and non-tolerable incidents, and striving to develop an increased understanding with the judicial authority, should contribute to constructing an appropriate level of transparency in ATM safety performance, while complying with legal requirements.

---

3   Some researchers even question safety reporting as the main safety feedback loop in high reliability organisations and in macro-technical systems.

4   www.eurocontrol.int/esp/public/standard_page/just_culture.html.

Therefore, the demand for safety reporting, intended to improve safety and risk management in ATM, is at the same time risking making the ATM profession appear closed, and possibly avoiding or unable to produce a 'simple' measure of safety. Such an image would conflict with society's expectation of transparency from safety-related professions. The question here is whether it can be counterbalanced with values carried by air traffic management such as safety, professionalism and protection of the public.

8.3.3   *ATM and Environmental Sustainability*

The sustainability of air transport, as with other overarching objectives like safety, is a shared objective that touches all actors in the industry. It translates concretely into all strategies and choices at different geographical scales, and different time scales. ATM positively contributes to air transport sustainability through cooperative, operational improvements such as energy efficiency and noise reduction.

Energy efficiency The contribution from air traffic management is essential to the improvement of flight efficiency, not only helping airlines to become more economical by consuming less fuel, but also reducing the environmental impact of aviation – and, usually, these go hand-in-hand. Better flight efficiency aims at reducing fuel consumption and gaseous emissions by a number of mechanisms. As we have seen in Chapter 2, airlines willing to optimise their energy efficiency need their routes to be validated by flow management functions, and must have their optimal route cleared by ATC. The 2005 Performance Review Report (EUROCONTROL, 2006a) reported that network inefficiencies cost airlines €1.4 billion each year. ATM can help to reduce inefficiencies of the European air transport network by up to 6 per cent, through:

* airspace design;
* flow management;
* Flexible Use of Airspace (FUA);
* ATC.

As an example, a EUROCONTROL study (Jelinek et al., 2002) has demonstrated the environmental gains provided by the introduction of RVSM in operations in 2002, introducing six new flight levels in European continental airspace, resulting in 310 000 tons of fuel saved in 2002. We have already discussed FUA elsewhere in this book, notably in Chapter 7, and it is clear that allowing civil aircraft to fly more direct routes will result in lower fuel burns.

Noise reduction Noise reduction by airlines at airports is another example of cooperative measures. In response to strict regulations to phase out noisy aircraft (as discussed in Chapter 5), noise reduction strategies involve the airframe and engine manufacturing industry providing quieter aircraft, airlines purchasing and operating them, and air traffic management introducing noise abatement procedures (e.g. Continuous Descent Approaches and Noise Preferential Routes (NPRs) – both discussed in Chapter 5).

Continuous Descent Approach (CDA) is an example of a noise abatement procedure developed at some large airports and being progressively implemented more widely in Europe, often with the help of EUROCONTROL. Its basic principle is to avoid (as far as safety constraints allow) both long, low-altitude noisy flight segments, and the levelling of aircraft, during airport approaches. CDA adoption requires the cooperation of pilots and controllers, their respective companies and managers, plus airport operations managers.

A recent study of attitudes to change in ATM, *Attitudes to Societal Demands in ATM Operations* (Cook and Tanner, 2005b), exploring how operational ATC staff, pilots and management have implemented CDAs in different airports in Europe, has shown that CDA approaches require strategic and tactical support, in day-to-day decisions, in order to work:

> The prompt to adopt a particular course of action may be determined by a macro-level objective (e.g. to increase capacity at a European level), or by a micro-level objective (e.g. to reduce noise complaints at a specific airport). Strategic trade-offs can be made for both macro-level and micro-level objectives, whereas tactical trade-offs are mostly made in the context of micro-level objectives. This Report suggests that there is some deferment of social responsibility from pilots and controllers to the 'system', as established at the strategic level, in which the airport may play a special role.

This is represented by Figure 8.1.

		Level of objectives	
		Micro	**Macro**
Level of trade-offs	**Strategic**	Building a new runway (public protest against airport development *versus* support to local economy from airport)	Increase of European capacity (public demand for travel *versus* national/ European legislation on air quality)
	Tactical	CDA for flight AB123 (public protest against noise *versus* early morning capacity constraint at airport)	X

Figure 8.1   Trade-off levels *versus* levels of objectives

Sustainability assessment As we have discussed, air traffic management may be involved in public consultation processes mainly around airports and under the flight paths in TMAs, where air transport projects have to comply with state regulations and, therefore, to the EC Directive on Strategic Environmental

Assessment (2001/42/EC). In addition, in the context of the Single European Sky, 'light' environmental impact assessment could also be part of the European ATM Master Plan in SESAR. The aim would be to consider, at the design stage, the possible socio-economic and environmental impacts of the changes brought by SESAR at the global network level and to prepare the eventual state-level assessments.

In such sustainability assessment processes, air traffic management has a role to play in proposing technical alternatives and potentially providing information to the public on the implications of various options. ATM can contribute in the form of the design of airspace, arrival and departure routes, and operational procedures; in addition to traffic forecasts and impact studies. This implies that in order to be properly prepared, air traffic management needs to understand the needs and expectations of society, which may then become a full ATM stakeholder. This also implies a responsibility to communicate with the public about missions, objectives, constraints and impacts, and represents an opportunity to demonstrate transparency and professionalism.

8.4   Partnering ATM and Society

In the previous sections we have elaborated the challenges faced by air transport in the future and the challenges of societal factors associated with change in air traffic management. Now we will explore how ATM could possibly address those challenges. The mainly technical and operational ATM responses to societal challenges are entirely necessary. However, they may not be sufficient in the socio-political sense. Various 'soft' issues, pertaining to the relationship with several categories of players in society, are not yet fully addressed by ATM, possibly leading to a relative weakness of air traffic management on the political scene.

Whilst continuing its drive to improve performance, greater awareness of ATM's relationship with the public and more transparency will be required from the air traffic management community, in order to respond to the challenges of public image, safety transparency and sustainability. How should air traffic management behave in this new political landscape? How can it make itself understood, make allies instead of enemies, and be recognised for its professionalism? We will now turn our attention to three areas, to consider this evolving context, *viz.* ATM's:

- partners in the new business context;
- communication skills;
- ability to change.

Contributing to securing the social acceptability of air transport activities and of ATM's constraints, is a challenging objective for the air traffic management community.

8.4.1 ATM's Partners in the New Business Context

Firstly, we need to consider all the ATM stakeholders in the new business context defined by the continuous growth, the liberalisation of services and the increased power of customers. The aim is to understand their needs and expectations, and to be recognised in return. ATM stakeholders include air transport players on the one side, and society players on the other.

Within air transport, developing a broader understanding of trends in the market can result in improving the relationship between air traffic management and its professional environment. As an example, airlines and airports perform market research (both tracking and *ad hoc* surveys) to be in tune with their business environment. ATM managers would also benefit from anticipating changes in the air transport business. Changes in transport demand, for instance, are likely to affect all involved in air transport. In air traffic management, understanding transport demand, travel requirements, and the strategies of other (air) transport players in the market, would certainly help towards finding solutions adapted to future transport needs.

Anticipation can help any business or service to be prepared for, and responsive to, change – and to avoid developing defensive attitudes. Within ATM, this is likely to improve its image within the air transport sector, contributing to further developing an inclusion culture.

From the perspective of society, those engaged in ATM aiming to improve air transport sustainability would increasingly need to acknowledge and recognise the public as the wider stakeholders in air transport, namely: passengers, airport residents, sometimes victims (whose interests are mainly local), associations, local authorities, businesses, politicians, journalists, and citizens in the broader context (who may have local and global interests). Their needs should be considered when developing strategies in air traffic management, for example as part of corporate social responsibility. Such recognition is indeed fostered by the EU Strategic Environmental Assessment regulations, mentioned earlier. When ATM is involved in decision-making processes affecting the public (such as mediation, consultation, assessment, etc, as discussed earlier), such acknowledgment implies accommodating room for manoeuvre for constructing acceptable compromises.

It also means being prepared to possibly share operational decisions with these stakeholders, in the wider sense – informed laypeople, with societal values – where professional values of safety, efficiency and optimisation may have prevailed to the *exclusion* of broader considerations. This new attention to neophytes' opinions implies a cultural change in the air traffic management community, traditionally rather technical and not used to communicating its choices.

8.4.2 ATM's Communication Skills

The second area to consider is communication, for which trust and transparency are essential components. Communication is a two-way process involving both information provision (promoting transparency) and listening to others' needs (promoting trust).

Proactively communicating about its role, operations and constraints, would contribute to consolidating ATM's image to the public. Accurate information and explanations have to be developed about its operations, design of air routes and likely consequences for the surrounding communities, operational concepts, technological choices and management practices. Some of the ATC job characteristics – such as responsibility, team spirit or the highly technological environment – could be good drivers for reinforcing public awareness of air traffic management. Such information could form the basis for building and demonstrating *transparency*.

Demonstrating openness and attention to safety, to the population's needs and to the other air transport actors' needs, would contribute to building *trust* in the ATM community.

However, developing the ATM relationship with society has associated difficulties that should not be underestimated, besides the requirement of allocating proper resources in the long run to build transparency, and maintain accurate information and dialogue.

First of all, to build the foundations for the public to understand explanations from the ATM community, background communication could develop general awareness among the public and reinforce the image of air traffic management. On some occasions, ATM may need to go further and to be able to explain exceptions – which is difficult in a profession which does not currently know how to communicate on its nominal performance. Efforts should be devoted to understanding what is needed by the public to comprehend ATM and what information might be interesting to them.

Secondly, openness also carries some risks since it involves building trust between parties who do not always share the same values. As an example of contrasting values, the desire of the industry to continue to grow complies with the desire of population to travel, but contradicts the neighbourhood's environmental needs, above certain levels of traffic.

The case of safety transparency (as discussed earlier), which receives support, in principle, from national authorities, and at the same time raises a number of issues in operations, is interesting as it mobilises professional values (safety management principles) and societal values (legal provisions in some countries) which both concur to the same value of protection of the public.

8.4.3   *ATM's Ability to Change*

An important aspect of the relationship of air traffic management with the outside world is its own ability to adapt to pressures of external change. Expectations, uncertainties, and, in some cases, discomfort, are associated with the preparations for the profound changes ATM is undergoing, in particular, when they affect the social side of the profession. This industrial sector is used to effectively combating operational risk and uncertainties with robust solidarity and relative social stability. Developing awareness and understanding of these pressures for change is an enabler for the required adaptation, since this makes it easier to accommodate anticipated changes. Learning from success stories in air traffic management and other industries is also a way of facilitating adaptation. Uncertainties created by future changes in ATM include:

- possible increasing automation of controller tasks: routine tasks (as well as certain separation tasks) could be carried out increasingly by computers, and controllers could be only intervening at crucial times (e.g. to guarantee separation, or resolve system degradation). The fear is that such changes would move the controller's job more towards management and/or supervision work, and involve the discarding of existing and highly valued skills, plus the adoption of new (and perhaps less challenging) tasks. Research into new operational concepts tend to address human-centred automation and transition phases – including experimentation and simulation as an important part. Unfortunately, previous attempts at automation have created a poor image for controllers, and this mediates and bounds controllers' expectations. Controllers are told that the automation is to remove 'human error', but, paradoxically, they may see many system errors (for example, programming faults in display systems or incompatibilities between systems that should be interoperable). Such errors they are then expected to accommodate. Moreover, critically, automation can and has created classes of new errors that can make the system brittle in certain respects;
- personnel moves in the context of the Single European Sky: the implementation of the SES has created many opportunities for change in ATM, with the potential of social impacts and uncertainties on working conditions for many personnel on issues such as: licences for air traffic controllers and other ATM personnel; changing management in the ATM industry; and reorganisation of air navigation service provision into Functional Airspace Blocks (see also Chapter 7). As an example, the creation of a Community air traffic controller licence aims to standardise controllers' qualifications across Europe and to facilitate controllers' mobility across operational centres – bringing both opportunities and threats. In response to these fears, the EU launched a number of studies to analyse technical aspects of airspace and service provision. It also launched a social dialogue on air traffic management, aiming to develop the social dimension of the SES, and to promote EU-wide agreements on work organisation and on working conditions.

Interesting research on change already exists, in particular in Sweden (Arvidsson et al., 2005) on the role the organisational culture plays in enabling change in ATM. We also refer again to the study *Attitudes to Societal Demands in ATM Operations* (Cook and Tanner, 2005b), which explores how controllers, pilots, and management, have implemented operational changes in a societal context – a discussion which has been further explored in a recent paper (Cook et al., 2007). This study draws on the Seven Stages of Change model, developed by the EU TAPESTRY project, originally devised to better understand the public's attitude to change in respect of assessing a cross-section of European initiatives focusing on promoting environmentally friendly travel behaviour (Tyler and Cook, 2004). The ATM study, addressing in its first stage the introduction of new noise abatement procedures, shows very diverse levels of assimilation by aviation staff of the need to respond to societal needs across Europe in day-to-day operations, and some transfer of social responsibility from pilots and controllers to 'the system'.

Such studies facilitate greater understanding of success factors in the air traffic management change process, highlighting the influence of staff participation and buy-in on such success. Nor should the importance of leadership in change be underestimated – although this is more difficult to study through research. Management schools and consultants build some of their educational programmes on change on experience gathered in a wide range of industrial sectors: such experience could help air traffic management learn from other industries.

8.5   Conclusions

The air transport industry must now, and for the foreseeable future, come to terms with a new reality, namely the pressure on air transport to assume its external costs, the growing resistance to 'uncontrolled' air transport growth, the demand for citizen participation and zero-risk tolerance. Should the industry continue to ignore this new reality and run the risk of being unprepared? Or should it anticipate the possible changes?

The image of air transport is still (mostly) strong and positive. Although its shine is starting to wear thin in places, this provides all those involved in the air transport sector – air traffic management included – with a good basis for addressing the negative trends in a proactive manner. Quoting R. Webster (of easyJet) at the *ANAE Colloquium on Air Transport and the Energy Challenge*, in December, 2006:

> Air transport needs to be collectively seen to behave as 'responsible citizens'.

ATM must make its full contribution to air transport sustainability – safety being an essential element of such sustainability. The recognised technical and operational competency of the ATM community is mobilised and ready to make a proactive contribution to operational sustainability improvements. In addition, the ATM community needs to learn how to communicate, listen to, and incorporate the needs of, society in its working practices, to demonstrate its ability to understand and to respond to society's requests. In order to build a relationship based on trust with the public and maintain its recognised legitimacy in society, in addition to the current ATM performance-driven improvement programmes, cultural changes need to be implemented in air traffic management to cope with societal demands.

Research can help us to understand changes in the environment of ATM and to take the right decisions when defining adaptation strategies. In European ATM, in line with the European community policy, EUROCONTROL has become progressively more active in taking account of environmental impacts, and has been conducting wider research into the social and economic issues surrounding air transport. The prospective studies of the EUROCONTROL Experimental Centre Strategy Unit aim to provide the air transport industry and policy-makers with an enhanced understanding of these societal and economic aspects and how they are evolving, paving the way for better-informed decisions. The 7th European Union Framework Programme for research, adopted by the European Parliament and Council at the end of 2006, includes provisions for developing a wider understanding of the socio-economic context in which air transport will have to evolve in the next decades.

Conclusions and a Look Ahead

Andrew Cook
University of Westminster

Drawing succinct conclusions on the prospects of European ATM is not at first sight a particularly tractable task. Where does one start with a range of issues as diverse as those examined in the preceding chapters? In an attempt to marshal some thoughts together, three core themes will be briefly explored here: automation, data integration, society and climate change.

Capacity has for a long time been the single greatest challenge facing European ATM. Now, increasingly, environmental issues are rising on the political and social agendas in various European states (albeit at different speeds) and, relatively more recently, in the US. Some of the solutions to these problems will be provided by new technologies, such as lighter (composite) aircraft and improved engines, but the time pressure is severe and ATM should be able to deliver results more quickly than the time required for most of these other technological changes to take place.

Such change will occur in the increasingly challenging context of rapid air traffic growth, promoted by greater liberalisation. The EU–US 'open skies' agreement is now (finally) set to come into force at the end of March 2008. Such liberalisation is expected to particularly boost traffic for countries such as Spain and Ireland, these not having previously benefited from such rights. It will enable all EU-based carriers to fly from any point in the EU to any airport in the US, and *vice versa*. Whilst it will grant US airlines (further) fifth-freedom rights within the EU, the reverse is not true: EU carriers are not thus enabled to land at one US airport and fly on to another. For this, and several other reasons, the agreement has met with a mixed response and the focus of attention for many airlines is already on the second-stage deal, although the existing agreement itself took several years to agree.

Most developments in the wider air transport industry, and ATM in particular, seem to take place over painfully protracted periods. Changes in airspace structures and the associated procedures take a long time to design, and usually longer to implement. Even a relatively straightforward new air traffic control system can take the best part of a decade to introduce, and several of the wider-ranging projects have taken up to 15 years.

Aircraft which will not be flying until after 2020 are already on order but many will require retrofits of new technologies by the time they are in service. New airport terminal facilities designed to accommodate the needs of a particular airline, or alliance network, are sometimes redundant before they even open.

With regard to SESAR (the Single European Sky ATM Research programme), although its final (deployment) phase ends in 2020, its scope will reach well beyond that. The programme, jointly funded by EUROCONTROL and the European Commission, critically includes a further-reaching vision than 2020, foreseeing the need for built-in flexibility in the evolution of ATM it seeks to establish. The same may be said of its US counterpart, NextGen (Next Generation Air Transportation System) which is led by the Joint Planning and Development Office.

Notable is the emerging prominence of several newer themes in such programmes. NextGen has taken the 'gate-to-gate' notion of air travel forward to a 'curb-to-curb' concept, which aims to include the increasingly challenging problems of airport security, now felt especially keenly on both sides of the North Atlantic. Indeed, the EU–US 'open skies' agreement will also establish a joint committee to harmonise standards between the EU and the US on security and safety, in addition to competition.

ATM takes place over great distances, often crossing continents and oceans. In addition to growth, a related challenge is more effective data communications over these distances and timescales, not only between different stakeholders but even within the large operational entities involved. It is not always evident to all divisions of an airline, such as dispatchers and flight planners, who could be hundreds of miles apart, which flight is the one with the most critical delay to reduce. Equally, it is not always clear to different sectors of the same Air Navigation Service Provider (ANSP), which aircraft should be performing a Continuous Descent Approach or what local holding measures are in place. Nor even can it be taken for granted that one sector knows how the adjacent sector wishes to have its traffic delivered. Controllers do not always have all the information they want on every flight that enters their airspace, and there is certainly room for improvement in both the timing and supply of data to allow controllers to detect and resolve evolving conflicts earlier – automation will certainly have a role to play here.

Although Air Traffic Control (ATC) works within very stringent rules, it is often not appreciated how much flexibility and decision-making lies in the hands of controllers. When dealing with commercial flights, although controllers must keep the aircraft within controlled airspace (which, by definition, is airspace under the direct control of ATC) they can use the entirety of such airspace to maintain their number one priority: safety. In fact – and in some contrast to the airline perspective – more important than knowing the precise trajectory of the aircraft *per se*, ATC is more concerned with *compliance*, in order to avoid and resolve conflicts. Planning, monitoring and controlling the evolution of air traffic is a very labour-intensive task. Automation will play a key part in extracting ever more capacity from limited resources, particularly human resources.

These then are the themes we will now explore a little further in closing. They are clearly interconnected and impact on all areas of ATM, but for automation we will discuss the controller's viewpoint, for data integration we will focus more on the airline perspective, and regarding climate change, we will broaden the discourse to the societal level.

Automation

Automation has promised much to ATC for a long time but has rather under-delivered. Real progress has been slow; many processes and supporting technologies (including data integration, which we will take a look at in the next section) have changed relatively little in this respect. However, by no means a panacea, automation may well present another set of problems, notably the well-known risk of swapping human errors for computer errors. These may be more difficult for the controller to rectify, as their causes may be far from transparent: buried in code, or resulting from interoperability problems. Nevertheless, although it seems very unlikely that the controller will be replaced in anything like the foreseeable future, especially in complex sectors, it is expected that the nature of the task will slowly start to change.

Currently, the controller has three fundamental types of task, which can be described as planning, monitoring and communicating. The planning function requires the controller to constantly revise his strategy for managing the evolving traffic in his sector, taking into account both the traffic he already has under his control and new traffic expected. The traffic then needs to be monitored for conformity with that plan and for conflict resolution. This is often an immensely cognitively demanding task, and several attempts at automation have severely underestimated this complexity. The distinct layer of tasks associated with communication are those such as establishing an aircraft on frequency, transferring frequencies and performing handovers, verifying read-backs, and dealing with other pilot requests.

Three fundamental stages of automation can be envisaged: facilitation of communication (e.g. without any delegation of control); decision-support tools (perhaps with a partial delegation of control) and full automation (where the controller would only intervene in cases of various types of system failure).

Automation's next major advance will probably be with regard to looking after the communication layer of tasks, thus freeing up more controller attention and time for the higher-level tasks of planning and monitoring. This raises several questions. Firstly, how will the removal of one set of tasks affect the ability to perform the others? One analogy is with the latest generation of cars with cruise control, lane departure warning systems and satellite navigation with dynamically integrated traffic information. Here it is possible for the driver to become somewhat disassociated from the actual task of driving, as lower level tasks are removed, which previously required concentration and engagement. The role thus becomes generically more passive.

Secondly, we have the issue of intermediate levels of automation and the type of human-machine interfaces that could exist between such applications and the controller. There are very delicate balances of design which need to be considered here. Clearly, as a safety-critical system, it must have a fail-safe *modus operandi*. This, however, raises the classical problem that too many false alarms engender mistrust at best, delayed attention to truly critical situations at worst. For this level of automation to be useful, the application must at least take some of the more routine decision-making away from the controller without barraging him with streams of messages and warnings requiring attention. In this latter case, the controller is in a worse situation than before, for example by suddenly having to respond to a set of particular conflicts, without having been an active agent in the evolution of the traffic pattern up to that moment.

Under the scenario of full automation, the controller finds himself in more of a supervisory role, intervening in instances of partial system failure (e.g. responding to requests relating to conflicts which the system is not 'authorised' to solve, possibly requiring verbal negotiation with an adjacent sector) or, worse, a more fundamental degradation of control. In such latter cases, it is unlikely that the controller will be required to 'step in' to the situation and resolve it. In part, this may be because the skills of the controller will have evolved away from those of the present-day, cognitive monitoring skills, and partly because the very appreciation of how to resolve conflicting traffic patterns is dependent on active participation in their development, not just being suddenly presented with a screen full of blips.

Rather, future system recovery is more likely to be achieved through a different kind of mechanism, perhaps by an automated delegation of control of given aircraft to one or more stand-by centres, with inputs from 'controllers' (who may all be called 'supervisors' by this time) as to how this delegation should be handled. In the future context of such a paradigm shift in operations, purely manual fallback seems very unlikely and there certainly will not be time to re-write code during the failure event! Some admix of technological substitution and procedural recovery (as in present-day go-arounds) seems more likely.

A critical factor very often overlooked in this type of change process is indeed the human factors component: how to effectively manage such change from the controllers' point of view. Of course, such advances will affect pilots too, and aircraft will have their own associated advances in technology and automation, although we have focused on the controller perspective here. Such obstacles which need addressing are the poor image that certain failures in automation have left controllers and pilots with in the past, negative connotations of loss of control and/or professional responsibility, and how to manage the *intermediate* solutions.

It will be interesting to see how SESAR and NextGen manage to meet these needs, particularly in the transitional period. Currently, there is a great deficit of research projects addressing these issues. Nevertheless, although this section has been somewhat introspective with respect to the potential problems associated with automation, it is appropriate to close on a note stressing the enormous potential benefits which may be realised through their introduction, which makes it worth spending time and research effort on getting this right.

Data Integration

Fragmentation is a common theme in European ATM, as is the associated problem of data integration. However, relatively recent initiatives in the grand scheme of decades of evolution of European ATM, such as the establishment of

EUROCONTROL, combined with some degree of technological harmonisation, have led to a position which is looking increasingly positive. A new operations room at EUROCONTROL's Central Flow Management Unit (CFMU) opened in July 2007. It has two important features. Firstly, the processing of initial flight plans, airspace data management and flow management now all occur in the same (rather large) room. Secondly, it has one radar map for the whole of Europe. This is both an iconic step forward, as well as a technological one. The vision is one of greater data sharing between all stakeholders – airlines, airports and ANSPs – not restricted by national boundaries and contributing to the optimisation of the use of airspace.

However, even today, the CFMU picture is not always perfect. For example, it may not be aware when local tactical flow measures are applied (e.g. en-route and/or holding at airports). Furthermore, the number of airports participating in message exchange trials with CFMU is still very limited and this needs to be addressed, considering the key role (capacity-constrained) airports play in the European ATM process. CFMU's objective of true management of *capacity*, in addition to flows, has yet to be fully realised.

From the current airline perspective, if an aircraft is flying from Helsinki to Lisbon, and the airline wishes to track its position in real time in its flight watch tool, usually in order to update/manage its Estimated Time of Arrival (ETA), then it has to pay for feeds from a number of different suppliers (e.g. NATS). These feeds are mostly based on secondary radar data. Although CFMU provides a European-level feed to the ANSPs, there is currently no single, consolidated source of such data for airline tools.

This data situation in Europe, for airlines, is thus still somewhat behind the single feed available from the FAA in the United States, where there is, of course, the distinct advantage of a single ANSP. Even this provision is set to improve, as the US will have an extensive deployment of ADS-B as part of its NextGen programme. More on this in a moment, but as the FAA puts it: 'Automatic Dependent Surveillance Broadcast (ADS-B) is, quite simply, the future of air traffic control.'

There are, meanwhile, other solutions to monitoring for the airlines in Europe, for example by calculating ETAs based on known take-off times and airborne positional updates transmitted by datalink from the aircraft, then using flight plan and meteorological data. ETAs calculated by the cockpit's Flight Management System are also used, again transmitted to the ground by datalink. The reliability of these calculations is always highest when the full flight path is known, although this is often not the case, due to uncertainties associated with future ATC instructions, especially with regard to arrival management.

Integrated European data provision for the airlines should, however, change in the future and be enhanced by the widespread introduction of surveillance technologies such as ADS-B. By transmitting data to ground stations on aircraft position, status and trajectory, data which are then relayed to airline and flow-management tools on the ground, ADS-B will make an increasing contribution to true 4D trajectory management. This is a crucial component of Collaborative Decision Making, from departure slots to Required Times of Arrival. Of note here, is that ACARE (Advisory Council for Aeronautics Research in Europe) comments that each trajectory should not necessarily be optimal for each particular flight but should be environmentally optimised for the whole network. Whilst some fear that 'trajectory contracts' such as these will reduce flexibility too much, others are convinced that this will be outweighed by overall system efficiencies, even with *advantages* in flexibility, for example through slot swapping. In any case, the widespread use of 4D trajectory management in Europe is likely to be quite some time away yet and will need to build on closer coordination between ATC, ATM and the airlines.

To support such progress, particularly in the context of an emerging Single European Sky, and as commented upon above, we also need to ensure that controllers have timely provision of all necessary information, including what airspace and which routes are available. This depends on reliable data integration with military cells (from the strategic phase through to the tactical) and will help controllers to offer aircraft-preferred trajectories to the maximum extent possible. As EUROCONTROL's Performance Review Commission has stated: 'capacity and flight efficiency must be addressed in a balanced way, so that the total cost to users, comprising the marginal cost of capacity and cost of route extension, is minimised', saying that a performance target should be adopted for flight efficiency as a matter of urgency.

Although datalink technologies have been around for quite some time now and have already been used for various types of ATC clearances, the technology is still very much under-exploited. Not only can it confer the benefits described above, but it also alleviates controller radiotelephony workload, which is currently a critical determinant of sector capacities. Looking further still towards the future, air-to-air datalink and surveillance technologies (such as ADS-B) also have the capability of contributing to the delegation of separation to aircraft, thus further reducing controller workload.

The technology and infrastructure to generate and supply many of the currently unmet data needs outlined here already exist. It is now more a question of coordinating these activities both politically and physically, plus obtaining the necessary certifications, before we can extract the key benefits: improved capacity management, lower delays, and reduced emissions. Endorsing this need for increasing data sharing and commonality, as part of SESAR's interoperability objectives, common platforms for (radar) data exchange are being developed and will soon be emerging under EU legislation.

Society and Climate Change

A particular aspect of the evolution of air transport which is all-too-frequently overlooked is the societal one. Difficult questions remain unanswered on this issue, questions which are not going to go away. Even within the context of the relative political and social cohesion of the EU, it poses diverse challenges. How will government, society and industry balance the objectives of greater access to air travel for the less mature European markets, such as the recent Accession countries, with the opposing demand to reduce the impact of aviation on climate change, for example? Thinking at governmental levels is not always as joined-up as it should be, with conflicting messages relating to airport and capacity expansion on the one hand, and the need to curb CO_2 emissions on the other.

Arguments are presented that the emergence of the low-cost airlines has made international travel more accessible to the poorer in society, but evidence suggests that this has actually helped middle-income people fly more often, rather than improving equitability. As demand for air travel increases, the conflict with pressures to tackle the issue of climate change will be thrown into increasingly sharp relief. The economic and social benefits of air travel are clear; the alternatives to non-short-haul air travel are less clear. Whilst some short-haul flights may be substituted by growing high-speed rail networks and intermodal solutions, whilst we can heat our homes more efficiently, produce electricity more cleanly and improve our recycling behaviour, the medium-term options for most air travel are more of a challenge. Since allowing aviation to become a net purchaser of CO_2 trading permits produces an increase in the climate impact of each permit (due to the associated effects such as NO_x emissions), the trade-off is not a simple one. At a local level, around airports, we somehow have to weigh the effects of local impacts in terms of noise, against global impacts in terms of CO_2, which are sometimes in conflict, for example through the use of longer approach and departure routes to reduce noise footprints.

Indeed, airports act as a focal point for these potential conflicts of demand. They serve a sometimes strangely dichotomous role, both promoting their own business interests along with those of their primary business customer, the airline, and, at the same time, safeguarding the 'public' interest. Increasingly, airports have come to focus on the social management of their activities, not only as a result of societal influences through protest groups and the media, but also of existing and impending legislation regarding their environmental impact. It is somewhat curious how airlines have, by and large, escaped such repeated and direct action: there are not many protest groups outside airline offices. The airport will be the focal point of increasingly difficult decisions ahead: if there is not a concomitant investment in airport expansion, to complement increases in airborne capacity, the latter will suffer from a diminishing return as the result of continued capacity constraints on the ground. There is little point pouring vast sums into ATM if there is not going to be sufficient supporting airport infrastructure.

By early 2008, SESAR's Definition Phase (2005–2008) will deliver a European ATM Master Plan based on the foreseen future aviation requirements. To quote from its website: 'SESAR focuses on serving the needs of society by delivering an air traffic management system for the future. SESAR will provide the means to make change happen'. When the Definition Phase was launched, EU Transport Commissioner, Jacques Barrot, identified the parallel objectives of SESAR as: to increase safety by a factor of ten; to triple ATM capacity; to reduce the environmental impact by 10 per cent per flight and to halve unit costs. Crudely taking the capacity increase and environmental impact reduction together, this still means that this impact would increase by a factor of 2.7. It is expected that the

European ATM Master Plan will consider the socioeconomic and environmental impacts of the changes presented by SESAR – it is keenly awaited.

Successful airlines have proven themselves pretty adept at responding to, and anticipating, market pressures – constantly adapting their product. In such a competitive market, they have had no choice. In a new era of marketing tactics, we can now see many airlines very keen to demonstrate their 'Green' credentials, although many of the statements made are somewhat relativistic, for example comparing their emissions with those of a worse airline! However, we have to bear in mind the context of aviation emissions. Even in regions where the contribution to CO_2 emissions is high, such as in the UK, it is still considerably less than one tenth of the total from the combustion of fossil fuels, and around five times less than that, globally. It is true that this share is expected to increase, but one cannot help but feel that the aviation industry is perhaps too much presented as the climate change ogre. There have, however, been a few successful cases of positive media relationships between the aviation industry and the media. Scandinavia has long-since been at the fore of environmental action, and its largest home-based airline, SAS, has achieved a great media success in promoting its new 'Green Approaches' into Stockholm Arlanda airport, whereby aircraft fly Continuous Descent Approaches from the Top of Descent, right onto the tarmac, thus significantly reducing emissions and noise. Certainly a step in a more positive direction, in more ways than one.

~*~

In closing, we have explored some intimately interconnected themes. Automation and data integration are mutually dependent. Together, they will contribute to a paradigm shift in ATM: one which will allow far more efficient management of flight trajectories and promote the reduction of unnecessary emissions, thus helping to deliver the growth demanded by the public. There are undoubtedly some real challenges ahead – but there are some very exciting opportunities, too. Whilst steps towards improved data integration are well underway, and improved efficiencies through automation are starting to look more realisable, facing the challenge of societal conflicts and climate change has hardly even begun.

Appendix

European ATM: a Compilation of Web-Based Information Resources

Graham Tanner
University of Westminster

This section is comprised of a range of useful ATM web-based information resources. These have been categorised to assist the reader, although some of the information spans more than one category.

Websites with data or documents available for downloading are identified by the symbol '🗈'. At the time of publication, the websites listed provided the information/services as indicated. Amendments will be made to the online version of this section, which will be maintained at:

www.wmin.ac.uk/atmbook

The password required is the last word of Chapter 8.

General information sources (organisations, trade bodies, mailing lists, link portals and interest groups)

EUROCONTROL. *European Organisation for the Safety of Air Navigation – development, coordination and planning for implementation of pan-European ATM* 🗈: www.eurocontrol.int/
European Aviation Safety Agency (EASA). *European Union agency for safety and environmental protection in civil aviation* 🗈: www.easa.eu.int/
European Civil Aviation Conference (ECAC). *European intergovernmental civil aviation organisation* 🗈: www.ecac-ceac.org/
International Civil Aviation Organisation (ICAO). *Global forum for civil aviation* 🗈: www.icao.int/
International Civil Aviation Organisation (ICAO) European and North Atlantic (EUR/NAT) Office. *ICAO European and NAT regional office* 🗈: www.paris.icao.int/
Joint Aviation Authorities (JAA). *European organisation for the civil aviation regulatory authorities – cooperation in development/implementation of common safety regulatory standards and procedures* 🗈: www.jaa.nl/

Advisory Council for Aeronautics Research in Europe (ACARE). *European organisation for the planning of aeronautic research programmes* 🗐: www. acare4europe.org/

European Commission Directorate-General for Energy and Transport – Air Transport. *European air transport policies* 🗐: http://ec.europa.eu/transport/ air_portal/index_en.htm

Federal Aviation Administration (FAA). *US aviation regulator* 🗐: www.faa.gov/

EUROCONTROL – Institute of Air Navigation Services (IANS). *Air Traffic Management training* 🗐: www.eurocontrol.int/ians/

EUROCONTROL – TrainingZone. *Air Traffic Management training [registration required]* 🗐: http://elearning.eurocontrol.int/

AeroSpace and Defence Industries Association of Europe (ASD). *Trade association for European aeronautics, space, defence and security industries* 🗐: www.asd-europe.org/

Airport Operators Association (AOA). *Trade association for UK airports* 🗐: www. aoa.org.uk/

Airports Council International (ACI). *Trade association for airports around the world* 🗐: www.airports.org/

Airports Council International (ACI) – Europe. *Trade association for airports around Europe* 🗐: www.aci-europe.org/

Association of European Airlines (AEA). *Trade association for the European airline industry* 🗐: www.aea.be/

British Air Transport Association (BATA). *Trade association for the UK airline industry* 🗐: www.bata.uk.com/

British Airline Pilots' Association (BALPA). *UK flight crew association* 🗐: www. balpa.org.uk/

Civil Air Navigation Services Organisation (CANSO). *Global trade association for Air Navigation Service Providers (ANSPs)* 🗐: www.canso.org/

European Federation of Airline Dispatchers Associations (EUFALDA). *European flight dispatchers association*: www.eufalda.org/

European Low Fares Airline Association (ELFAA). *Trade association for the European low fare airline industry* 🗐: www.elfaa.com/

European Regions Airline Association (ERA). *Trade association for the European regional airline industry* 🗐: www.eraa.org/

European Union Airport Coordinators Association (EUACA). *Trade association for the European airport coordinators*: www.euaca.org/

Guild Of Air Traffic Control Officers (GATCO). *UK civil and military air traffic controller association*: www.gatco.org/

International Air Carrier Association (IACA). *Trade association for the leisure airline industry* 🗐: www.iaca.be/

International Air Transport Association (IATA). *Air transport industry trade body* 🗐: www.iata.org/

International Federation of Air Line Pilots' Associations (IFALPA). *Global trade association for pilots* 🗐: www.ifalpa.org/

International Federation of Air Traffic Controllers' Associations (IFATCA). *Global trade association for air traffic controllers* 🗐: www.ifatca.org/

Society of British Aerospace Companies (SBAC). *Trade association for UK civil air transport, aerospace, defence and space companies (includes British Airports Group)* 🖹 : www.sbac.co.uk/

EUROCONTROL – AIS AGORA. *Aeronautical information forum [registration required]*: www.eurocontrol.int/aisagora/

Flight Safety Foundation. *Aviation safety forum* 🖹 : www.flightsafety.org/

International Federation of Airworthiness (IFA). *Aviation safety forum* 🖹 : www. ifairworthy.com/

Airline Group of the International Federation of Operational Research Societies (AGIFORS). *Airline industry operational research society* 🖹 : www.agifors. org/

Royal Aeronautical Society (RAeS). *Aeronautical institution* 🖹 : www.raes.org. uk/

Jane's Information Group. *News and information* 🖹 : www.janes.com/

Airbus – My Airbus. *Airbus activities newsletter [subscription required]*: www. airbus.com/en/myairbus/subscription/

ATC Network. *Aviation forum and newsletter (monthly) [subscription required]*: www.atc-network.com/

Aviation Information Resources (AIR). *Aviation newsletter (twice-monthly) [subscription required]*: www.airlineupdate.com/

Aviation Interactive. *Aviation newsletter (daily) [subscription required]*: www. aviationindustrygroup.com/index.cfm?pg=173

Boeing Media. *Boeing activities newsletter (daily) [subscription required]*: http:// boeingmedia.com/mailing_list/

EUROCONTROL – Experimental Centre (EEC) News. *EEC activities newsletter (two/three per year) [subscription required]*: www.eurocontrol.int/eec/public/ standard_page/EEC_News_All_Issues.html

Aerospace Technology. *Collection of online aerospace industry resources* 🖹 : www. aerospace-technology.com/

Airline Contact Information. *Collection of airline links and contact details*: www. airlinecontact.info/

Airport Technology. *Collection of online airport industry resources* 🖹 : www. airport-technology.com/

Cranfield University AERADE Portal. *Collection of online aerospace and defence resources*: http://aerade.cranfield.ac.uk/

Landings. *Collection of online aviation resources*: www.landings.com/

PilotPointer.com. *Collection of online aviation resources*: www.pilotpointer.com/

Air Transport Action Group (ATAG). *Aviation industry interest group* 🖹 : www. atag.org/

Aviation Environment Federation (AEF). *Environmental interest group – aviation* 🖹 : www.aef.org.uk/

Enviro.aero. *Environmental interest group – aviation (part of Air Transport Action Group)* 🖹 : www.enviro.aero/

European Union Against Aircraft Nuisances (UECNA). *Environmental interest group – airports*: www.uecna.eu/

Fly On Track. *Airspace infringements awareness campaign* 📖: www.flyontrack.
co.uk/
GreenSkies. *Environmental interest group – aviation* 📖: www.greenskies.org/
NATS Level Best Campaign. *'Level bust' awareness campaign* 📖: www.levelbust.
com/
Sustainable Aviation. *UK Aviation sustainable development strategy* 📖: www.
sustainableaviation.co.uk/

ATM projects (current pan-European research projects and programmes)

Air Traffic Alliance. *Industry partnership led by EADS, Airbus and Thales –
evolution of ATM within the Single European Sky initiative* 📖: www.airtraffic-
alliance.com/
**Community Research and Development Information Service (CORDIS) – Air
Transport**. *EU – current air transport research projects*: http://cordis.europa.
eu/transport/src/air.htm
EUROCONTROL – Innovative Research (INO). *Innovative research and
development – current projects and INO workshops* 📖: www.eurocontrol.int/
eec/public/standard_page/INO.html
EUROCONTROL – Single European Sky (SES). *EU, EUROCONTROL and
ICAO Initiative – to reorganise European airspace and air navigation* 📖: www.
eurocontrol.int/ses/
EUROCONTROL – Single European Sky ATM Research Programme (SESAR).
*European programme (2005-2020) consisting of ATM stakeholders (civil,
military, legislators, industry, operators, users, ground and airborne) in support
of the Single European Sky initiative (formerly known as SESAME)* 📖: www.
eurocontrol.int/sesar/
European Collaborative Decision Making (CDM) Portal. *IATA, EUROCONTROL
and ACI Project – airport and regional collaborative decision making* 📖: www.
euro-cdm.org/
**European Commission Directorate-General for Energy and Transport – Single
European Sky (SES)**. *EU, EUROCONTROL and ICAO Initiative – to
reorganise European airspace and air navigation* 📖: http://ec.europa.eu/
transport/air_portal/traffic_management/ses/index_en.htm
SESAR Consortium. *Stakeholder consortium – airports, airlines, ANSPs and the
supply industry* 📖: www.sesar-consortium.aero/
ATC Maastricht. *Conference and exhibition*: www.atcmaastricht.com/
European Air Traffic Management Research and Development Symposium.
EUROCONTROL organised – bi-annual symposium programme 📖: http://
atmsymposium.eurocontrol.fr/
USA/Europe Air Traffic Management Research and Development Seminars. *FAA
and EUROCONTROL organised – bi-annual seminar programme* 📖: http://
atmseminar.eurocontrol.fr/

Understanding codes (encoding and decoding IATA/ICAO airport, airline and other aviation codes)

AeroTransport Data Bank. *Aviation resource – IATA/ICAO codes (aircraft type, airline, airport), links [subscription required]*: www.aerotransport.org/
Airline Codes Website. *Aviation resource – IATA/ICAO codes (aircraft type, airline, airport, callsigns), links*: www.airlinecodes.co.uk/
A-Z World Airports. *Airport resource – IATA/ICAO codes, links*: www.azworldairports.com/
IATA – Airline Members. *Airline database – IATA/ICAO codes*: www1.iata.org/membership/airline_members.htm
ICAO – Aircraft Type Designators. *Aircraft resource – ICAO codes (ICAO Doc 8643)*: www.icao.int/anb/ais/8643/
ICAO – Designators and Indicators. *Aircraft/airport resource – ICAO codes (ICAO Doc 7910, ICAO Doc 8585 and ICAO Doc 8643, hosted by EUROCONTROL)*: www.eurocontrol.int/icaoref/
ICAO – EUR/NAT Regional Database. *Aviation resource – ICAO codes (designated points and route designators, hosted by EUROCONTROL)* 📄: www.eurocontrol.int/icard/
World Aeronautical Database. *Airport/navaid resource – ICAO codes*: www.worldaerodata.com/

Flight planning and navigation sources (aeronautical information and charts, civil aviation authorities, ATFM messages, slot coordination services and flight planning tools)

AeroPlanner.com. *Flight planning resources [subscription required]* 📄: www.aeroplanner.com/
AirNav. *US aviation information – airports, navaids and fixes* 📄: www.airnav.com/
CANSO – Links to ANSPs. *Links to air navigation service provider websites*: www.canso.org/Canso/Web/links/Air+Navigation+Service+Providers/
EUROCONTROL – @is online. *Links to aeronautical information websites – e.g. access to various Aeronautical Information Packages (AIPs)*: www.eurocontrol.int/aim/public/standard_page/ais_online.html
EUROCONTROL – Aeronautical Information Management (AIM). *European aeronautical information* 📄: www.eurocontrol.int/aim/
EUROCONTROL – Airspace and Navigation. *European airspace – classification, procedures, flexible use (FUA), civil-military coordination* 📄: www.eurocontrol.int/airspace/
EUROCONTROL – European AIS Database (EAD). *European database of aeronautical information – e.g. access to various European Aeronautical Information Packages (AIPs) [registration required]* 📄: www.ead.eurocontrol.int/

EUROCONTROL – Navigation Domain. *European navigation strategy – e.g. Reduced Vertical Separation Minimum (RVSM) and Precision Area Navigation (P-RNAV)* 🗎: www.ecacnav.com/

Federal Aviation Administration (FAA) Air Traffic Control System Command Center (ATCSCC). *Access to various FAA services – e.g. flight delay information at US airports and Aviation Information System (AIS)* 🗎: www.fly.faa.gov/

ICAO – Links to CAAs. *Links to civil aviation authority websites:* www.icao.int/icao/en/m_links.html

Jeppesen. *Flight navigation and planning services [subscription required]* 🗎: www.jeppesen.com/

SAS Flight Dispatch (SFD). *SAS flight planning and other services:* www.sasflightdispatch.com/

Scandinavian Flight Operations. *SAS flight operations information – e.g. datalink:* www.sasflightops.com/

ICAO North Atlantic Programme Coordination Office (NAT PCO). *Information for North Atlantic Region ATC* 🗎: www.nat-pco.org/

EUROCONTROL – Cartography Service. *European Airspace Management Planning and Sectorisation charts* 🗎: www.eurocontrol.int/carto/

EUROCONTROL – SkyView2. *European aeronautical information Geographical Information System (GIS) application [registration required]* 🗎: www.eurocontrol.int/geoaeronet/public/standard_page/skyview_overview.html

Great Circle Mapper. *Plot great circle/geodesic paths using airport codes or coordinates:* http://gc.kls2.com/

Aeronautical Radio Incorporated (ARINC). *Air transport industry communications service provider* 🗎: www.arinc.com/

Société Internationale de Télécommunications Aéronautiques (SITA). *Air transport industry communications service provider* 🗎: www.sita.aero/

Airport Coordination Limited (ACL). *Airport slot allocation and slot coordination services* 🗎: www.acl-uk.org/

EUROCONTROL – Central Flow Management Unit (CFMU). *Provision of European ATFM service within area of responsibility of participating states* 🗎: www.cfmu.eurocontrol.int/

EUROCONTROL – Central Flow Management Unit (CFMU) – Public Applications. *Access to various CFMU services/messages – e.g. IFPS Validation System (IFPUV), Route Availability Document (RAD), ATFM Information Messages (AIM) and Conditional Route Availability Messages (CRAM)* 🗎: www.cfmu.eurocontrol.int/chmi_public/ciahome.jsp

EUROCONTROL – Central Flow Management Unit (CFMU) – Taxi Times. *European airports taxiing times for each runway* 🗎: www.cfmu.eurocontrol.int/cfmu/opsd/public/standard_page/operational_ services_taxitimes.html

Flughafenkoordination Deutschland (FHKD). *German airport slot allocation and slot coordination services – with links to other slot coordinators* 🗎: www.fhkd.org/cms/

European Aeronautical Group (EAG). *RODOS – suite of flight planning tools:* www.euronautical.com/

Lufthansa Systems Aeronautics (LSA) [formerly known as Lido]. *Lido Operations Center (Lido OC) – suite of flight planning and dispatch tools*: www.lido.net/

Navtech. *DispatchPro – suite of flight planning tools and dispatch tools* 🖹: www. navtechinc.com/

Information, data and statistical sources (air transport statistics, forecasts, aircraft fleet databases, environmental information, modelling data and manufacturers)

Air4casts. *Air passenger forecasts* 🖹: www.air4casts.com/

Airclaims. *Aviation survey and loss adjusting services – e.g. CASE database (commercial aircraft valuations and market information)*: www.airclaims. co.uk/

AVITAS. *Aviation financial and operational data – e.g. Jet BlueBook (commercial aircraft valuations) and GOAT Book (Global Outlook for Air Transportation)* 🖹: www.avitas.com/

Bureau of Transportation Statistics (BTS). *US Government agency – transport statistics* 🖹: www.bts.gov/

Civil Aviation Authority (CAA) – Aviation Statistics. *UK airport, airline and punctuality statistics* 🖹: www.caa.co.uk/statistics/

Department for Transport (DfT) – Aviation Directorate. *UK Government department – aviation statistics and information* 🖹: www.dft.gov.uk/aviation/

Encyclopedia of the Atmospheric Environment (EAE). *Manchester Metropolitan University – atmospheric information* 🖹: www.ace.mmu.ac.uk/eae/

EUROCONTROL – eCoda. *Central Office for Delay Analysis (CODA) – air transport delay reports* 🖹: www.eurocontrol.int/eCoda/

EUROCONTROL – STATFOR. *Air traffic statistics and short-, medium- and long-term forecasts* 🖹: www.eurocontrol.int/statfor/

FlightOntime.info. *Airline punctuality statistics – at major UK airports* 🖹: www. flightontime.info/

Heathrow Airwatch. *Heathrow air quality information* 🖹: www.heathrowairwatch. org.uk/

ICAO – ICAO Data. *Air transport industry statistical data [subscription required]* 🖹: www.icaodata.com/

National Atmospheric Emissions Inventory (NAEI). *UK emissions information* 🖹: www.naei.org.uk/

National Statistics (ONS). *UK Government agency – air transport statistics* 🖹: www.statistics.gov.uk/CCI/nscl.asp?ID=7905

Official Airline Guide (OAG) – data. *Airline schedules and travel information products [subscription required]* 🖹: www.oagdata.com/

The Aircraft Value Analysis Company. *Aviation values and lease rates [subscription required]*: www.aircraft-values.co.uk/

Airport Handling Services. *Information and company directory of airport ground handling services* 🖹: www.ahs1000.com/

Aviation Environmental Best Practice Database. *Aviation industry initiatives intended to reduce environmental impact [registration required]*: www.iata.org/ebpdb/

Boeing – Airport Noise Regulations. *Airport information – e.g. noise regulations and available runways* 🖹 : www.boeing.com/commercial/noise/

CAST/ICAO Common Taxonomy Team. *Aviation common taxonomies – e.g. aircraft, engine and phase of flight* 🖹 : www.intlaviationstandards.org/

Civil Aviation Authority (CAA) – Publications. *UK civil aviation publications* 🖹 : www.caa.co.uk/publications/search.asp

IATA – Ground Handling Council (IGHC) – Directory. *Company directory of airport ground handling services*: www.iata.org/ighc/public/search.aspx

IATA – Passenger Intelligence Services (PaxIS) Database. *Airline passenger ticket database*: www.pax-is.com/

ICAO – Annexes to the Convention on International Civil Aviation. *Booklet of annexes 1 to 18 (PDF format)* 🖹 : www.icao.int/icaonet/anx/info/annexes_booklet_en.pdf

ICAO – Civil Aviation and the Environment. *Aircraft noise and emissions* 🖹 : www.icao.int/icao/en/env/

Airfleets. *Aircraft information – e.g. production lists and current fleets* 🖹 : www.airfleets.net/

Airliners.net. *Aircraft information – e.g. development history and data* 🖹 : www.airliners.net/

Civil Aviation Authority (CAA) – G-INFO Aircraft Register Database. *UK civil aircraft registration database*: www.caa.co.uk/ginfo/

EUROCONTROL – Aircraft Noise and Performance (ANP) Database. *Aircraft noise modelling database [registration required]* 🖹 : www.aircraftnoisemodel.org/

EUROCONTROL – Base of Aircraft DAta (BADA). *Aircraft performance database* 🖹 : www.eurocontrol.fr/projects/bada/

FAA – Flight Standards Service. *US civil aircraft registration database* 🖹 : http://registry.faa.gov/

ICAO – Aircraft Engine Emissions Databank. *Aircraft exhaust emissions database (ICAO Doc 9646, hosted by Civil Aviation Authority)* 🖹 : www.caa.co.uk/default.aspx?categoryid=702&pagetype=90

ICAO – NoisedB. *Aircraft noise certification database (hosted by Direction Générale de l'Aviation Civile)* 🖹 : http://noisedb.stac.aviation-civile.gouv.fr/

A.S.Yakovlev Design Bureau (Yakovlev). *Aircraft manufacturer*: www.yak.ru/ENG/

Airbus. *Aircraft manufacturer* 🖹 : www.airbus.com/

ATR. *Aircraft manufacturer* 🖹 : www.atraircraft.com/

BAE Systems – Regional Aircraft. *Aircraft manufacturer* 🖹 : www.regional-services.com/

Boeing. *Aircraft manufacturer* 🖹 : www.boeing.com/

Bombardier Aerospace. *Aircraft manufacturer* 🖹 : www.aero.bombardier.com/

Embraer. *Aircraft manufacturer* 🖹 : www.embraercommercialjets.com/

Ilyushin. *Aircraft manufacturer*: www.ilyushin.org/eng/

Sukhoi. *Aircraft manufacturer*: www.sukhoi.org/eng/

Tupolev. *Aircraft manufacturer*: www.tupolev.ru/english/

Glossary

ABI	Advance Boundary Information
ACARE	Advisory Council for Aeronautics Research in Europe
ACC	Area Control Centre
ACI	Airports Council International
ACT	Activation Message
ADC	Air Data Computer
ADEXP	ATS Data Exchange Presentation
ADF	Automatic Direction Finder
ADP	Aéroports de Paris
ADP	ATFM Daily Plan
ADS	Automatic Dependent Surveillance
ADS-B	Automatic Dependent Surveillance – Broadcast
AEA	Association of European Airlines
AENA	Aeropuertos Españoles y Navegación Aérea (Spain)
AFIL	Air Filed Flight Plan
AFTN	Aeronautical Fixed Telecommunications Network
AGIFORS	Airline Group of the International Federation of Operational Research Societies
AIC	Aeronautical Information Circular
AIM	Aeronautical Information Management
AIM	Air Traffic Flow Management Information Message
AIP	Aeronautical Information Publication
AIRAC	Aeronautical Information, Regulation and Control
AIS	Aeronautical Information Services
AMHS	Air Traffic Services Message Handling System
AMSL	Above Mean Sea Level
ANM	Air Traffic Flow Management Notification Message
ANS	Air Navigation Service
ANSP	Air Navigation Service Provider
AO	Aircraft Operator
AOA	Airport Operators Association
AOWIR	Aircraft Operator 'What-If' Re-route
APP	APP Approach Centre/Control
APU	Auxiliary Power Unit
ARINC	Aeronautical Radio Incorporated
ARTCC	Air Route Traffic Control Center
ASD	AeroSpace and Defence Industries Association of Europe
ASM	Air Space Management
ATAG	Air Transport Action Group
ATC	Air Traffic Control

ATCU	Air Traffic Control Unit
ATFCM	Air Traffic Flow and Capacity Management
ATFM	Air Traffic Flow Management
ATM	Air Traffic Management
ATO	Actual Time Over
ATS	Air Traffic Services
ATSU	Air Traffic Service Unit
B-RNAV	Basic Area Navigation
CAA	Civil Aviation Authority (UK)
CAMSIM	Canadian Management Simulator
CAPAN	Capacity Analyser
CASA	Computer Assisted Slot Allocation
CBA	Cross Border Area
CCR	Centre de Contrôle Régional
CDA	Continuous Descent Approach
CDM	Collaborative Decision Making
CDR	Conditional Route
CEN	European Committee for Standardisation
CENA	Centre d'Etudes de la Navigation Aérienne (France)
CENELAC	European Committee for Electrotechnical Standardisation
CFMU	Central Flow Management Unit
CHG	Modification Message
CHMI	CFMU Human Machine Interface
CIA	CFMU Internet Application
CIAO	CFMU Human Machine Interface for Aircraft Operators
CIDIN	Common ICAO Data Interchange Network
CIFLO	CFMU Human Machine Interface for Flow Management Positions
CIR	CFMU Interactive Reporting
CIREN	CFMU Human Machine Interface for ENV Coordinators
CNL	Cancellation Message
CNS	Communication, Navigation and Surveillance
CORDIS	Community Research and Development Information Service
CPDLC	Controller-Pilot Data Link Communications
CRCO	Central Route Charges Office
CRS	Computer Reservation System
CSF	Complexity Shaping Factor
CTOT	Calculated Take-Off Time
CVSM	Conventional Vertical Separation Minima
DCT	Direct
DFS	Deutsche Flugsicherung (Germany)
DLA	Delay Message
DME	Distance Measurement Equipment
DSNA	Direction des Services de Navigation Aérienne
DST	Decision Support Tool

EAD	European AIS Database
EAM	European Airspace Model
EASA	European Aviation Safety Agency
EATCHIP	European Air Traffic Control Harmonisation and Integration Programme
EATM	European Air Traffic Management
EATMP	European Air Traffic Management Programme
ECAC	European Civil Aviation Conference
ECMT	European Conference of Ministers of Transport
EEC	EUROCONTROL Experimental Centre
EET	Estimated Elapsed Time
EOBT	Estimated Off-Block Time
ESARR	EUROCONTROL Safety Regulatory Requirement
ETFMS	Enhanced Tactical Flow Management System
ETO	Estimated Time Over
ETOT	Estimated Take-Off Time
ETSI	European Telecommunications Standards Institute
EUROCAE	European Organisation for Civil Aviation Equipment
EUROCONTROL	European Organisation for the Safety of Air Navigation
FAA	Federal Aviation Administration (US)
FAB	Functional Airspace Block
FANS	Future Air Navigation Systems
FASTI	First ATC Support Tools Implementation
FDP	Flight Data Processor
FFP	Frequent Flyer Programme
FIR	Flight Information Region
FIS	Flight Information Service
FL	Flight Level
FLS	Flight Suspension Message
FMP	Flow Management Position
FMS	Flight Management System
FMU	Flow Management Unit
FPL	Filed Flight Plan
FTS	Fast-Time Simulation
FUA	Flexible Use of Airspace
GAT	General Air Traffic
GBAS	Ground Based Augmentation System
GCD	Great Circle Distance
GDP	Gross Domestic Product
GDS	Global Distribution System
GNSS	Global Navigation Satellite System
GPS	Global Positioning System
HMI	Human Machine Interface
IAF	Initial Approach Fix
IANS	Institute of Air Navigation Services
IATA	International Air Transport Association

ICAN	International Convention for Air Navigation
ICAO	International Civil Aviation Organisation
IFALPA	International Federation of Air Line Pilots' Associations
IFATCA	International Federation of Air Traffic Controllers' Association
IFPS	Integrated Initial Flight Plan Processing System
IFPUV	IFPS Validation System
IFPZ	IFPS Zone
IFR	Instrument Flight Rules
ILS	Instrument Landing System
INS	Inertial Navigation System
IPCC	Intergovernmental Panel on Climate Change
JAA	Joint Aviation Authorities
LAM	Logical Acknowledge Message
LAQ	Local Air Quality
LFV	Luftfartsverket (Sweden)
LTO	Landing-Take-Off
MAEVA	Master ATM European Validation Plan
MASPS	Minimum Aviation System Performance Specification
MCT	Minimum Connect Time
MEL	Minimum Equipment List
MET	Metrorological Service
MLIT	Ministry of Land, Infrastructure and Transport (Japan)
MONA	Monitoring Aid
MSA	Minimum Sector Altitude
MSL	Mean Sea Level
MTCD	Medium Term Conflict Detection
MTOW	Maximum Take-Off Weight
MVA	Minimum Vectoring Altitude
NAEI	National Atmospheric Emissions Inventory
NAT	North Atlantic
NATS	National Air Traffic Services (UK)
Nav Portugal	Navegação Aérea de Portugal
NDB	Non-Directional Beacon
NM	Nautical Mile
NOTAM	Notice to Airmen
NPR	Noise Preferential Route
OAG	Official Airline Guide
OAT	Operational Air Traffic
OLDI	On-Line Data Interchange
OTA	Observational Task Analysis
PC	Planning Controller/Planner
PIB	Pre-flight Information Bulletin
PITOT	Process-based Integrated PlaTform for Optimal use of analysis Techniques
PPM	Parts Per Million

PRC	Performance Review Commission
PRFPL	CFMU RPL Input Application
P-RNAV	Precision Area Navigation
PRR	Performance Review Report
PRU	Performance Review Unit
PUMA	Performance and Usability Modelling in ATM
RAD	Route Availability Document
RAIM	Receiver Autonomous Integrity Monitoring
RAMS	Re-Organized ATC Mathematical Simulator
RCA	Reduced Coordination Airspace
RCA	Remote Client Application
REA	Ready Message
RFI	Ready/Request For (direct) Improvement Message
RFP	Replacement Flight Plan Procedure
RNAV	Area Navigation
RNP	Required Navigation Performance
RPL	Repetitive Flight Plan
RRN	Re-routing Notification Message
RT	Radiotelephony
RTA	Remote Terminal Access
RTS	Real-Time Simulation
RVR	Runway Visual Range
RVSM	Reduced Vertical Separation Minima
SAM	Slot Allocation Message
SAR	System Analysis Recording
SDAT	Sector Design and Analysis Tool
SES	Single European Sky
SESAR	Single European Sky ATM Research Programme
SICTA	Sistemi Innovativi per il Controllo del Traffico Aereo (Italy)
SID	Standard Instrument Departure
SIMMOD	Airport and Airspace SIMulation MODel
SIP	Slot Improvement Proposal Message
SIS	Support Information System
SIT	Slot Issue Time
SITA	Société Internationale de Télécommunications Aéronautiques
SLC	Slot Requirement Cancellation Message
SMM	Slot Missed Message
SPA	Slot Improvement Proposal Acceptance Message
SRD	Standard Route Document
SRJ	Slot Improvement Proposal Rejection Message
SRM	Slot Revision Message
SSR	Secondary Surveillance Radar
STAR	Standard Terminal Arrival Route
SWM	SIP Wanted Message

SYSCO	System Supported Coordination
TAAM	Total Airspace Airport Modeller
TAS	True Air Speed
TC	Tactical Controller
TMA	Terminal Manoeuvring Area or Terminal Control Area
TRACON	Terminal Radar Approach Control (US)
TSA	Temporary Segregated Area
TWR	Tower Control Unit
UIR	Upper Flight Information Region
UNFCCC	United Nations Framework Convention on Climate Change
UTC	Coordinated Universal Time
VFR	Visual Flight Rules
VOR	VHF Omnidirectional Range

Bibliography

Advisory Council for Aeronautics Research in Europe (2002) *Strategic Research Agenda Volume 2: The challenge of the environment* (ACARE).

Advisory Council for Aeronautics Research in Europe (2003) *The Economic Impact of Air Transport on the European Economy Report* (ACARE), September 2003.

Advisory Council for Aeronautics Research in Europe (2004) *Strategic Research Agenda 2* (ACARE).

Air Transport Action Group (2005) *The Economic and Social Benefits of Air Transport* (ATAG) September 2005.

Airports Council International Europe (2004a) *Night Flight Restrictions*, 3rd edn (ACI Europe).

Airports Council International Europe (2004b) *The Social and Economic Impact of airports in Europe* (ACI Europe).

Alderighi, M. and Cento, A. (2004) 'European Airlines Conduct after September 11', *Journal of Air Transport Management*, 10, 97–107.

Arthur D. Little Limited (2000) *Study into the Potential Impact of Changes in Technology on the Development of Air Transport in the UK*, Final Report to Department of the Environment, Transport and Regions (DETR), DETR Contract No. PPAD 9/91/14.

Arvidsson, M., Johansson, C.R., Ek, A. and Akselsson, R. (2005) *Organisational Climate in Air Traffic Control – Innovative Preparedness for Implementation of New Technology and Organizational Development in a Rule Governed Organization*, Applied Ergonomics, Elsevier.

Association of European Airlines, AeroSpace and Defence Industries Association of Europe, European Express Association, European Business Aviation Association, European Regions Airline Association and The International Air Carrier Association (2005) *European Aviation Industry Joint Position Paper on Emissions Containment Policy*.

Association of European Airlines (2005) *AEA Yearbook 2005*, Association of European Airlines, Brussels.

Aviation Strategy (2001) 'Ryanair, Just Too Good a Negotiator', *Aviation Strategy*, July/August.

Ayral, M. (2003) 'La Communauté européenne et le transport aérien', *Petites affiches – les dossiers de l'Europe*, 30 January 2003.

Barrett, S. (2000) 'Airport Competition in the Deregulated European Aviation Market', *Journal of Air Transport Management*, 6, 13–27.

Bates, J., Polak, J., Jones, P. and Cook, A. (2001) 'The Valuation of Reliability for Personal Travel', *Transportation Research Part E* (37), 191–229.

Beatty, R., Hsu, R., Berry, L. and Rome, J. (1998) *Preliminary Evaluation of Flight Delay Propagation through an Airline Schedule*, 2nd USA/Europe Air Traffic Management R&D Seminar (Orlando, December 1998).

Beck, U. (1992) *Risk Society: Towards a new modernity*, London: Sage, 1992.

Belgocontrol (2007a) *Belgium and GD of Luxembourg Aeronautical Information Package*, ENR 3–2 Upper ATS routes, 28 September 2006, accessed May 2007.

Belgocontrol (2007b) *Belgium and GD of Luxembourg Aeronautical Information Package*, ENR 3-3 Area navigation (RNAV) routes, 15 March 2007, accessed May 2007.

Boeing (1999) *Airline Metric Concepts for Evaluating Air Traffic Service Performance*, Air Traffic Services Performance Focus Group.

Boeing (2006) *Current Market Outlook*, Boeing Commercial Airplanes, Seattle.

Boeing (2007) 737 Family: 737-800 Technical Characteristics web page, accessed May 2007. www.boeing.com/commercial/737family/pf/pf_800tech.html.

Bourgois, M., Cooper, M., Duong, V., Hjalmarsson, J., Lange, M. and Ynnerman, A. (2005) *Interactive and Immersive 3D Visualization for ATC*, ATM 2005 – 6th USA/Europe R&D Seminar.

Bråthen, S., Johansen, S. and Lian, JI. (2006) *An Inquiry into the Link between Air Transport and Employment in Norway*, European Transport Conference, 2006.

Bratu, S. and Barnhart, C. (2004), *An Analysis of Passenger Delays using Flight Operations and Passenger Booking Data*, Sloan Industry Studies Working Paper WP-2004-20.

Breivik, K. (2003) 'Enhanced Tactical Flow Management System: A clearer picture of the air traffic situation in Europe', *Skyway*, 7, 28, Spring.

Bristow, A., Wardman, M., Heaver, C., Murphy, P., Hume, K., Dimitriu, D., Plachinski, E., Hullah, P. and Elliff, T. (2003) *Attitudes Towards and Values of Aircraft Annoyance and Noise Nuisance – 5A Survey Report and Questionnaire*, EUROCONTROL Experimental Centre, EEC Note: EEC/SEE/2003/002.

Bristow, A., Wardman, M. and Hullah, P. (2005) *Further Analysis of the 5A Attitudinal and Stated Preference Data Sets*, EUROCONTROL Experimental Centre, EEC Note: EEC/SEE/2005/006.

Brooker, P. (2004) *The UK Aircraft Noise Index Study: 20 years on*, Proceedings of the Institute of Acoustics, 26, 20–30.

Brooker, P. (2006) 'Civil Aircraft Design Priorities: Air quality? Climate change? Noise?', *The Aeronautical Journal*, 110, 517–32.

Bruntland, G.H. (1987) *Our Common Future, World Commission on Environment and Development* (WCED), Oxford: Oxford University.

CAST Consortium (1998) *Consequences of future ATM systems for air traffic controller Selection and Training, WP1: Current and future ATM systems*, Contract No.: AI-97-SC.2029, European Commission.

Civil Aviation Authority (2005a) *Noise Exposure Contours for Heathrow Airport 2004*, CAA Environmental Research and Consultancy Department, Report 0501.

Civil Aviation Authority (2005b) *Noise Exposure Contours for Stansted Airport 2004*, CAA Environmental Research and Consultancy Department, Report 0503.

Civil Aviation Authority (2005c) *UK Regional Air Services – A Study by the Civil Aviation Authority*, CAA CAP 754, London.

Civil Aviation Authority (2006a) *No-frills Carriers: Revolution or evolution? A Study by the Civil Aviation Authority*, CAA CAP 770, London, The Stationery Office.

Civil Aviation Authority (2006b) *UK Airline/Airport Statistics, Annual*. CAA, London.

Civil Aviation Authority (2006c), *Manual of Air Traffic Services Part 1*, CAP 493 Part 1 (Amdt. 69, 28 April 2006), Safety Regulation Group (UK) CAA, ISBN 0 11790 587 9.

Civil Aviation Authority (2006d) *The UK Flight Planning Guide*, CAA CAP 694, 2nd edn, July 2006.

Civil Aviation Authority (2007a) *UK Aeronautical Information Package*, ENR 1-9 Air Traffic Flow Management (ATFM), CAA, 23 November 2006, accessed May 2007.

Civil Aviation Authority (2007b) *UK Aeronautical Information Package*, ENR 1-10 Flight Planning, 23 November 2006, CAA, accessed May 2007.

Civil Aviation Authority (2007c) *UK Aeronautical Information Package*, GEN 1-5 Aircraft Instruments, Equipment and Flight Documents, 10 May 2007, CAA, accessed May 2007.

Civil Aviation Authority (2007d) *UK Aeronautical Information Package*, AD 2-EGLL-6-5 London Heathrow Dover/Detling Standard Departure Chart, 06 July 2006, CAA, accessed May 2007.

CAP 493 Part 1 – see CAA (2006c) [CAP: (UK) Civil Aviation Publication].

CAP 694 – see CAA (2006d) [CAP: (UK) Civil Aviation Publication].

CONSAVE (2005) *CONSAVE 2050: Constrained Scenarios on Aviation and Emissions*.

Cook, A., Tanner, G. and Anderson, S. (2004) *Evaluating the True Cost to Airlines of One Minute of Airborne or Ground Delay*, University of Westminster, Report commissioned by the Performance Review Commission (EUROCONTROL).

Cook, A. and Tanner, G. (2005a) *'Citizens' Study: Results of European focus groups examining public perceptions of air transport growth and ATM*, EUROCONTROL Experimental Centre, EEC Note: EEC/SEE/2005/013.

Cook, A. and Tanner, G. (2005b) *Attitudes to Societal Demands in ATM Operations: Introduction of a B-CDA trial at Manchester, Bucharest and Stockholm*, EUROCONTROL Experimental Centre, EEC Note: EEC/SEE/2005/018.

Cook, A., Tanner, G., Pilon, N. and Joyce, A. (2007) *A Psychosocial Approach to Understanding Pilot and Controller Acceptance of Change in ATM, Based On Three CDA Case Studies*, 7th USA/Europe Air Traffic Management R&D Seminar, Barcelona, Spain, 2007.

Cooper, A. (2005) *The Economic Catalytic Effects of Air Transport in Europe*, EUROCONTROL Experimental Centre, EEC Note: EEC/SEE/2005/004.

Defra (2007) *e-Digest Statistics about: Climate Change. UK Emissions of Greenhouse Gases. Estimated emissions of carbon dioxide (CO_2) by IPCC source category, type of fuel and end user: 1970–2005.* Retrieved from: http://www.defra.gov.uk/environment/statistics/globatmos/download/xls/gatb05.xls.

Dennis, N. (2005) 'Industry Consolidation and Future Network Structures in Europe', *Journal of Air Transport Management*, 11, 175–83.

Dennis, N. (2007 forthcoming) 'Competition and Change in the Long-Haul Markets from Europe', *Journal of Air Transportation*, 12, 2.

Deutsche Flugsicherung (2007) *German Aeronautical Information Package*, AD 2-EDDF-3-1-2 Frankfurt Main Runway 07/25 Standard Arrival Chart, 12 April 2007, DFS, accessed May 2007.

Doc 4444 – see ICAO (2001).

Doganis, R. (2006) *The Airline Business*, 2nd edn, Routledge, London.

Dowling, A.P. and Hynes, T. (2006) 'Towards a Silent Aircraft', *The Aeronautical Journal*, 110, 487–94.

Ehrmanntraut, R. (2005) *Performance Parameters of Speed Control & the Potential of Lateral Offset*, EUROCONTROL Experimental Centre, Note 22/05.

EUROCONTROL (1987) *Future ATS System Concept Description*.

EUROCONTROL (1991) *European ATC Harmonization And Integration Programme (EATCHIP) – Report Phase 1*.

EUROCONTROL Experimental Centre (1995) *RAMS system overview document, Model Based Simulations Sub-Division*.

EUROCONTROL Experimental Centre (2005) *Airport Local Air Quality studies – ALAQS*, EEC Note: EEC/SEE/2005/016.

EUROCONTROL (1998) *Air Traffic Management Strategy for the Years 2000+*, vols 1 and 2, Brussels, Belgium.

EUROCONTROL (1999a) *Model Simulation of Bulgarian Airspace*, EEC Note 4/99, SIM-F-E1 (F7), ATM Operational & Simulation Expertise Group.

EUROCONTROL (1999b) *Navigation Strategy for ECAC*, NAV.ET1.ST16-001 edn 2.1.

EUROCONTROL (2000) *Reporting and Assessment of Safety Occurrences in ATM (ESARR 2)*, Version 2.0, Safety Regulation Commission, November 2000.

EUROCONTROL (2001a) *Status of Civil-Military Co-ordination in Air Traffic Management*.

EUROCONTROL (2001b) *EUROCONTROL Standard Document for ATS Data Exchange Presentation (ADEXP)*, DPS.ET1.ST09-STD-01-01, 2nd edn (V2.1), December 2001. www.eurocontrol.int/eatm/public/standard_page/library_standards_doc.html.

EUROCONTROL (2001c) *Repetitive Flight Plan (RPL) Input Application Users Guide (PRFPL)*, 3rd edn (V3.0), 12 November 2001.

EUROCONTROL (2002) *The EUROCONTROL Concept of the Flexible Use of Airspace*, October 2002.

EUROCONTROL (2003a) *Performance Review Report 6: An Assessment of Air Traffic Management in Europe during the Calendar Year 2002* (PRR 6), May 2003.

EUROCONTROL (2003b) *Flight Delay Propagation – Synthesis of the Study, M3 Systems*, EEC Note No 18/03, EUROCONTROL Experimental Centre.

EUROCONTROL (2004a) *Challenges to Growth 2004 Report* (CTG04).

EUROCONTROL (2004b) *Information to Accompany the Service Agreements – Acceptable Behaviour*, 1st edn (V1.0), October 2004.

EUROCONTROL (2004c) *Impact of Automation on Future Controller Skill Requirements and a Framework for their Prediction.*

EUROCONTROL (2005a) *Annual Report 2004*, July 2005.

EUROCONTROL (2005b) *Aeronautical Information Lexicon*, EATM, 21 September 2005.

EUROCONTROL (2006a) *Performance Review Report 2005: An assessment of Air Traffic Management in Europe during the calendar year 2005* (PRR 2005), April 2006.

EUROCONTROL (2006b) *Long-Term Forecast: IFR Flight Movements 2006– 2025*, EUROCONTROL.

EUROCONTROL (2006c) *FASTI OFG, Operational Concept*, EUROCONTROL.

EUROCONTROL (2006d) *Evaluation of the Impact of the Single European Sky Initiative on ATM Performance*, Performance Review Commission, 21 December 2006.

EUROCONTROL (2006e) *Long-Term Forecast: 2006 Report (*LTF06), EUROCONTROL.

EUROCONTROL (2006f) *Basic CFMU Handbook: General and CFMU systems*, 11th edn (V11.0), 02 May 2006.

EUROCONTROL (2006g) *Basic CFMU Handbook: ATFCM Users' Manual*, 11th edn (V11.0), 02 May 2006.

EUROCONTROL (2006h) *The Impact of Fragmentation in European ATM/ CNS*, Helios Economics and Policy Services, Report commissioned by the Performance Review Commission, April 2006.

EUROCONTROL (2006i) CFMU Flight Planning web page, accessed May 2007. www.cfmu.eurocontrol.int/cfmu/public/standard_page/about_flightplanning. html.

EUROCONTROL (2006j) *European AIS Database – Your Gateway to the World's Largest Aeronautical Information Repository*, June 2006.

EUROCONTROL (2006k) *Supplement to the CFMU Handbook: Route Availability Document Users Manual*, 1st edn (V1.0), 31 July 2006.

EUROCONTROL (2006m), *Complexity Metrics for ANSP Benchmarking Analysis, ACE Working Group on Complexity*, Report commissioned by the Performance Review Commission.

EUROCONTROL (2007a) *EATM: European Air Traffic Management, Our Commitment to Performance.*

EUROCONTROL (2007b) *Basic CFMU Handbook: IFPS Users' Manual*, 11th edn (V11.2), 30 March 2007.

EUROCONTROL (2007c) *CFMU Areas of Operation*, AIRAC 0701, 18 January 2007.

EUROCONTROL (2007d) CFMU Applications Access web page, accessed May 2007. www.cfmu.eurocontrol.int/cfmu/public/standard_page/services_and_support_ service_access_applications.html.

EUROCONTROL (2007e) European Aeronautical Fixed Service (AFS) web page, accessed May 2007. www.eurocontrol.int/cidin/public/standard_page/ European_AFS.html.

EUROCONTROL (2007f) *Route Restrictions Through Germany – ED: Annex ED*, AIRAC: 10 May 2007.

EUROCONTROL (2007g) *Route Availability Document Appendix 3: City-pair Level Capping*, AIRAC: 10 May 2007.

EUROCONTROL (2007h) *Route Availability Document Appendix 4: DCT Limits*, AIRAC: 10 May 2007.

EUROCONTROL (2007i) *ATFCM and Capacity Report 2006*, 1st edn (V1.0), February 2007.

EUROCONTROL (2007j) *Performance Review Report 2006: An assessment of Air Traffic Management in Europe during the calendar year 2006* (PRR 2006), May 2007.

European Commission (2001a) *European Transport Policy for 2010: Time to decide*, White Paper, COM (2001) 370.

European Commission (2001b) *Single European Sky*, Report of the High-Level Group, 2001.

European Commission (2002) *A Single European Sky: Broadening horizons for air travel*, Energy and Transport DG.

European Commission (2006) *Mid-term review of the Transport White Paper: Keep Europe Moving – Sustainable Mobility for our Continent*, European Commission Communication, June 2006.

European Conference of Ministers of Transport (2005) *Airports as multimodal interchange nodes* (ECMT) Report of Round Table 126, March 2003, ISBN 92-821-0339-0, ECMT, Paris 2005.

Faburel, G. (2005) 'L'espace aérien et les aéroports: l'épreuve des territoires', in Faburel (ed.) *Les Cahiers Scientifiques du Transport*, 47.

Faburel, G. (2006) *Les effets des trafics aériens autour des aéroports franciliens – Séminaires d'échanges sur les connaissances scientifiques et sur les indicateurs pour l'aide à la décision*, vols 1 and 2.

Fassert, C. (2005) *A Comparative Approach on Safety in Several Air Navigation Services Providers: Which role for Culture?*, presented at the Air Transport Research Society Conference, Rio de Janeiro, 2005.

Forster, P.Md.F., Shine, K.P. and Stuber, N. (2006) 'It is Premature to Include Non-Co$_2$ Effects of Aviation in Emission Trading Schemes', *Atmospheric Environment*, 40, 1117–21.

Fournie, A. (2005) 'Getting There on Time: The background to air traffic punctuality', *Skyway*, 9, 39, Winter.

Francis, G., Dennis, N., Ison, S. and Humphreys, I. (2007) 'The Transferability of the Low-Cost Model to Long-Haul Airline Operations', *Tourism Management*, 28, 2, 391–8.

Gawron, V.J., Scheflett, S.G. and Miller, J.C. (1989) 'Measures of In-Flight Workload', in Jensen, R.S. (ed.), *Aviation Psychology*, Brookfield, VA, Gower, 240–87.

Graham, A. and Dennis, N. (2007) 'Airport Traffic and Financial Performance: A UK and Ireland case study', *Journal of Transport Geography*, 15, 161–71.

Graham, B. and Vowles, T.M. (2006) 'Carriers within Carriers: A Strategic Response to Low-Cost Airline Competition', *Transport Reviews*, 26, 1, 105–26.

Guibert, S. and Guichard, L. (2005) *Paradigm SHIFT – Dual Airspace concept assessment*, 4th EUROCONTROL Innovative Research Workshop, Bretigny, France.

Guichard, L. (2000) *Du strip papier au système sans strip: un long chemin à parcourir* [*'Digistrip' – a Possible Electronic Strip Rack of the Future*], Human Factor Base, EUROCONTROL, Paris.

Haines, M.M., Stansfeld, S.A., Job, R.F.S., Berglund, B. and Head, J. (2001) 'Chronic Aircraft Noise Exposure, Stress Responses, Mental Health and Cognitive Performance in School Children', *Psychological Medicine*, 31, 265–77.

Halpern, C. (2006) *Le Istanze politiche della decisione pubblica. Le mobilitazioni intorno all'aereoporto Charles de Gaulle di Parigi*, Vitale (a cura di), Partecipazione a rappresentanza nelle mobilitazioni locali, Milano 2006, Franco Angeli.

Hansen, M.M., Gillen, D. and Djafarian-Tehrani, R. (2001) 'Aviation Infrastructure Performance and Airline Cost: A statistical cost estimation approach', *Transportation Research Part E* (37), 1–23.

Hilburn, B. (2004) *Cognitive Complexity in Air Traffic Control: A literature review*, EEC Note 04/04.

Hopkin, V.D. (1995) *Human factors in air traffic control*, Taylor snf Francis, London.

Hubbert, M.K. (1956) *Nuclear Energy and the Fossil Fuels*, presented before the Spring Meeting of the Southern District, American Petroleum Institute, Plaza Hotel, San Antonio, Texas, 1956.

Hudgell, A.J. and Gingell, R.M. (2001) *Assessing the Capacity of Novel ATM Systems*, paper presented at the 4th USA/EUROPE Air Traffic Management R&D Seminar, Santa Fe, USA.

INECO Study (2005) *Study on the Functioning of the Internal Market – Part 2: Land-use planning and management*, November 2005.

Intergovernmental Panel on Climate Change (1999) *Aviation and the Global Atmosphere, a special Report of the Intergovernmental Panel on Climate Change Working Groups I and III* (IPCC), Cambridge University Press.

Intergovernmental Panel on Climate Change (2001) *Climate Change 2001, the Scientific Basis. Summary for policymakers and technical Summary of* the Intergovernmental Panel on Climate Change Working Group I (IPCC), Cambridge University Press.

International Air Transport Association (2007) IATA Jet Fuel Price Monitor, www.iata.org/whatwedo/economics/fuel_monitor/index.htm.

International Civil Aviation Organisation (1998) *Convention on International Civil Aviation – Annex 11*, ICAO (Montreal), 12th edition.

International Civil Aviation Organisation (2001) *The Procedures for Air Navigation Services – Air Traffic Management* (PANS-ATM), ICAO Doc 4444, 14th edn, November.

International Civil Aviation Organisation (2004) *Operational Opportunities to Minimize Fuel Use and Reduce Emissions*, ICAO Circular, 303-AN/176.

International Civil Aviation Organisation (2006) 'Annual Review of Civil Aviation 2005', *ICAO Journal*, 61, 6–42.

Jelinek, F., Carlier, S., Smith, J. and Quesne, A. (2002) *The EUR RVSM Implementation Project: Environmental Benefit Analysis*, EUROCONTROL Experimental Centre, EEC Note: EEC/SEE/2002/008.

Jorna, P.G.A.M. (1991) *Operator Workload as a Limiting Factor in complex Systems, Automation and Systems Issues*, ed. Wise, J.A., Hopkin, D.V. and Smith, M., NATO ASI series Vol. F733 Springer-Verlag, Berlin.

Jovanović, R. (2006) 'Effects on Airline Delay Costs of Delayed Passengers', MSc Research Dissertation, Transport Planning and Management, Transport Studies Group, University of Westminster, London.

Kim, B., Fleming, G., Balasubramanian, S., Malwitz, L., Lee, J., Waitz, I., Klima, K., Locke, M., Holsclaw, C., Morales, A., McQueen, E. and Gillette, W. (2005) *System for assessing Aviation's Global Emissions Global Aviation Emissions Inventories for 2000 through 2004*, Version 1.5, Federal Aviation Administration Office of Environment and Energy, FAA-EE-2005-02.

Kinchin, B. (2004) *A Synthesis of ATM Public Perception Surveys*, EURO-CONTROL Experimental Centre, EEC Note: EEC/SEE/2004/001.

Kok, W. (2004) *Facing the Challenge: The Lisbon strategy for growth and employment*, Report from the High Level Group chaired by Wim Kok.

Koutrakos, P. (2000) 'Is Article 297 EC a "Reserve of Sovereignty"?', 37 CML Rev. (2000), 1339.

Laudeman, I.V., Shelden, S.G., Branstrom, R. and Brasil, C.L. (1998) *Dynamic Density: An air traffic management metric*, NASA Technical Paper (NASA/TM-1998-112226), NASA Ames Research Center, Moffett Field, CA.

Lawton, T. (2002) *Cleared for Take-Off: Structure and Strategy in the low-fare airline business*, Ashgate, Aldershot.

Magill, S.A.N. (1997) *Trajectory Predictability and Frequency of Conflict-Avoiding Action*, presented at the CEAS 10th International Aerospace Conference, Amsterdam, The Netherlands.

Mahaud, P. (2004) *What Image of ATM? An Analysis of 2002–2003 European Press*, EUROCONTROL Experimental Centre, EEC Note: EEC/SEE/2004/002.

Mahaud, P. (2005) *The Safety of Air Traffic as seen by the Press*, EURO-CONTROL Experimental Centre, EEC Note: EEC/SEE/2005/007.

Mahaud, P. and Courty, G. (2007) *Air Transport growth in Europe as seen by the political actors (2000-2006)*, submitted to the European Transport Conference, 2007.

Majumdar, A. and Polak, J. (2001) *Estimating the Capacity of Europe's Airspace using a Simulation Model of Air Traffic Controller Workload*, Transportation Research Record, 1744, 30–43.

Majumdar, A. and Ochieng, W.Y. (2002) *The Factors Affecting Air Traffic Controller Workload: A multivariate analysis based upon simulation modelling of controller workload*, Transportation Research Record, 1788, 58–69.

Majumdar, A., Ochieng, W.Y., McAuley, G., Lenzi, J.M. and Lepadatu, C. (2004) 'The Factors Affecting Airspace Capacity in Europe: A Cross-Sectional Time-Series Analysis Using Simulated Controller Workload', *Journal of Navigation*, 57, 3, 385–405.

Majumdar, A., Ochieng, W.Y., Bentham, J. and Richards, M. (2005) 'En-route Sector Capacity Estimation Methodologies: An international survey', *Journal of Air Transport Management*, 11, 375–87.

Majumdar, A. and Ochieng, W.Y. (2007 forthcoming) *Air Traffic Control Complexity and Safety: A framework for sector design based upon controller interviews of complexity factors*, Transportation Research Records.

Mannstein, H., Spichtinger, P. and Gierens, K. (2005) 'A Note on How to Avoid Contrail Cirrus', *Transportation Research Part D*, Transport and Environment, 10, 421–6.

Marland, G., Boden, T.A. and Andres, R.J. (2007) *Global, Regional, and National CO_2 Emissions, in: Trends: A Compendium of Data on Global Change*, Carbon Dioxide Information Analysis Center, Oak Ridge National Laboratory, US Department of Energy, Oak Ridge, TN.

Metrot, F. (2006) *The Energy Dilemma: European Air Transport growth between the devil and the deep blue sea, the Air Transport Research Society conference*, Nagoya, 2006.

Michel, S. (1995) L'homme et la Machine, Hallwag, Switzerland.

Mogford, R.H., Guttman, J.A., Morrow, S.L. and Kopardekar, P. (1995) *The Complexity Construct in Air Traffic Control: A review and synthesis of the literature*, DOT/FAA/CT-TN92/22, Department of Transportation/Federal Aviation Administration Technical Center, Atlantic City, NJ.

Morrish, S.C. and Hamilton, R.T. (2002) 'Airline Alliances – Who Benefits?', *Journal of Air Transport Management*, 8, 401–7.

National Air Traffic Services (2007) *UK and Ireland Standard Route Document*, NATS, AIRAC: 10 May 2007.

National Atmospheric Emission Inventory (2006) www.naei.org.uk.

Nichols, W.K. and Kunz, M. (1999) 'Hubbing on Time', *Airline Business*, August.

Nuutinen, H. (2005) 'MAXjet and Eos: Prospects for the all-business class concept', *Aviation Strategy*, October, 3–6.

Nuutinen, H. (2007) 'Delta Air Lines: From bankruptcy to industry-leading financials?' *Aviation Strategy*, April, 2–10.

Odoni, A.R., Bowman, J., Delahaye, D., Deyst, J.J., Feron, E., Hansman, R.J., Khan, K., Kuchar, J.K., Pujet, N. and Simpson, R.W. (1997), 'Existing and Required Modelling Capabilities for Evaluating ATM Systems and Concepts', Final Report, Modeling Research Under NASA/AATT, International Center For Air Transportation, Massachusetts Institute of Technology, Cambridge, MA.

Penner, J.E., Lister, D.H., Griggs, D.J., Dokken, D.J. and McFarland, M. (eds) (1999) *Aviation and the Global Atmosphere: A Special Report of Intergovernmental Panel on Climate Change Working Groups I and III*, Cambridge University Press, Cambridge.

Revuelta, J. (2000) MAEVA, *A Master ATM European VAlidation Plan: Overview of the MAEVA Project*, EC DG-TREN Transport Programme (2.3.1/2) Contract 2000-AM.10011, European Commission.

Rolls Royce (2003) *Powering a Better World*, Environment Report 2003.

Rypdal, K. (2000) *Aircraft Emissions in Background Papers IPCC Expert Meetings on Good Practice Guidance and Uncertainty Management in National Greenhouse Gas Inventories*.

Sausen, R., Isaksen, I., Grewe, V., Hauglustaine, D., Lee, D.S., Myhre, G., Kohler, M.O., Pitari, G., Schumann, U., Stordal, F. and Zerefos, C.S. (2005) 'Aviation Radiative Forcing in 2000: An update on IPCC (1999)', *Meteorol*, Zeitschrift, 14, 555–61.

SESAR Consortium (2006a) *Paving the Way for the Implementation of the Single European Sky*.

SESAR Consortium (2006b) *Air Transport Framework, the Performance Target SESAR Definition Phase: Deliverable 2*, DLM-0607-001-02-00a.

Sillard, L., Vergne, F. and Desart, B. (2000) *TAAM Operational Evaluation*, EEC Report Number 351, EUROCONTROL Experimental Centre.

SITA (2006) SITA's history and milestones web page, accessed May 2007, www.sita.com/News_Centre/Corporate_profile/History/default.htm.

Skyguide (2001), *ABC des services de la navigation aérienne*, Geneva.

Skyway editorial article (2005) 'Air Traffic Flow and Capacity Management: Optimising the efficiency of the ATM system', *Skyway*, 9, 39, Winter.

Stallen, P.J.M. and Compagne, H. (2006), 'Residential Preferences in Noise Annoyance Reduction: Individual exposure history matters much', University of Leiden, submitted to *Journal of Policy Analysis and Management*.

Stamp, R.G. (1990) *The Assessment of Airspace Capacity*, speaking notes to the Royal Aeronautical Society, London, UK.

Stamp, R.G. (1992) *The DORA TASK Method of Assessing ATC Sector Capacity – An Overview*, DORA Communication 8934, Issue 2, Civil Aviation Authority, London.

Suzuki, Y. (2000) 'The Relationship between On-Time Performance and Airline Market Share: A new approach', *Transportation Research Part E* (36), 139–54.

Sweetman, B. (2004) 'Bigger and Smarter', *Air Transport World*, May, 24–30.

Swyngedouw, E.(2000) *Elite Power, Global Forces, and the Political Economy of 'Glocal' Development*, The Oxford Handbook of Economic Geography, Oxford University Press, New York and Oxford.

Thomas, C. and Lever, M. (2003) 'Aircraft Noise, Community Relations and Stakeholder Involvement', in Upham, P., Maughan, J., Raper, D.W. and Thomas, C. (eds), *Towards Sustainable Aviation*, Earthscan, London, 97–112.

Tyler, S. and Cook, A. (2004) *Measuring the Effectiveness of Campaigns: Lessons for mobility management from the EU TAPESTRY project*, 8th European Conference on Mobility Management, Lyon, European Platform on Mobility Management, 2004.

US Energy Information Administration (2007), http://tonto.eia.doe.gov/dnav/pet/pet_pri_spt_s1_d.htm.

Van Houtte, B. (2000) 'Air Transport', in Geradin, D. (ed.), *The Liberalization of State Monopolies in the European Union and Beyond*, Kluwer, Amsterdam, 67–97.

Williams, V. and Noland, R.B. (2005) 'Variability of Contrail Formation Conditions and the Implications for Policies to Reduce the Climate Impacts of Aviation', *Transportation Research Part D: Transport and Environment*, 10, 269–80.

Williams, V. and Noland, R.B. (2006) 'Comparing the CO_2 Emissions and Contrail Formation from Short and Long Haul Air Traffic Routes from London Heathrow', *Environmental Science and Policy*, 9, 487–95.

Williams, V., Noland, R.B. and Toumi, R. (2002) 'Reducing the Climate Change Impacts of Aviation by Restricting Cruise Altitudes', *Transportation Research Part D: Transport and Environment*, 7, 451–64.

Wu, C.-L. and Caves, R.E. (2002) 'Towards the Optimisation of the Schedule Reliability of Aircraft Rotations', *Journal of Air Transport Management*, 8, 419–26.

Index